Homework Book, **Higher 2**

Delivering the Edexcel Specification

NEW GCSE MATHS

Edexcel Linear

Fully supports the 2010 GCSE Specification

Brian Speed ⬤ **Kevin Evans**
Trevor Senior ⬤ **Chris Pearce**

CONTENTS

INTRODUCTION

Welcome to Collins New GCSE Maths for Edexcel Linear Higher Homework Book 2.
This book follows the structure of the Edexcel Linear Higher Student Book 2.

Colour-coded grades

Know what target grade you are working at
and track your progress with the colour-coded
grade panels at the side of the page.

Use of calculators

Questions when you could use a calculator
are marked with a  icon.

Examples

Recap on methods you need by reading
through the examples before starting the
homework exercises.

Functional maths

Practise functional maths skills to see how
people use maths in everyday life. Look out
for practice questions marked (FM).

There are also extra functional maths and
problem-solving activities at the end of every
chapter to build and apply your skills.

New Assessment Objectives

Practise new parts of the curriculum
(Assessment Objectives AO2 and AO3) with
questions that assess your understanding
marked (AU) and questions that test if you can solve problems marked (PS). You will also
practise some questions that involve several steps and where you have to choose which
method to use; these also test AO2. There are also plenty of straightforward questions (AO1)
that test if you can do the maths.

Student Book CD-ROM

Remind yourself of the work covered in class with the Student Book in
electronic form on the CD-ROM. Insert the CD into your machine,
click 'Open a PDF file' and choose the chapter you need.

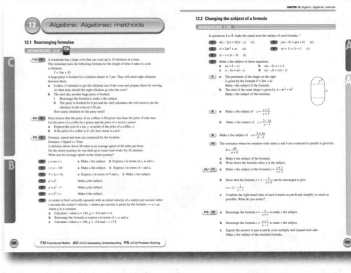

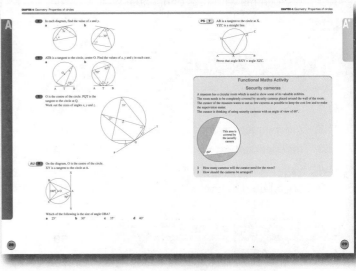

1.1 Basic calculations and using brackets

HOMEWORK 1A

Use your calculator to work out the following questions. Try to key in the calculation as one continuous set, without writing down any intermediate values.

1 Work out:
   a $(18 - 5) \times 360 \div 24$ b $360 - (180 \div 3)$

2 Work out:
   a $\frac{1}{2} \times (6.4 + 9.2) \times 3.6$ b $\frac{1}{2} \times (1.7 + 11.5) \times 7.3$

3 Work out the following and give your answers to one decimal place.
   a $\pi \times 7.8$ b $2 \times \pi \times 6.1$ c $\pi \times 10.2^2$
   d $\pi \times 1.9^2$

FM 4 A monthly travel ticket costs £61.60.
Karen usually spends £4.70 each day on travel.
How many days would she need to travel each month so that it would be cheaper for her to buy a monthly travel ticket?

AU 5 A teacher asked her class to work out: $\dfrac{3.1 + 5.2}{1.9 + 0.3}$

Alfie keyed in:

(3 . 1 + 5 . 2) ÷ 1 . 9 + 0 . 3 =

Becky keyed in:

3 . 1 + 5 . 2 ÷ (1 . 9 + 0 . 3) =

Chloe keyed in:

3 . 1 + 5 . 2 ÷ 1 . 9 + 0 . 3 =

Daniel keyed in:

(3 . 1 + 5 . 2) ÷ (1 . 9 + 0 . 3) =

They each rounded their answers to three decimal places.
Work out the answer that each of them got.
Who had the correct answer?

PS 6 £1 is equivalent to £1.14 Euros.
£1 is equivalent to 1.49 US dollars ($).
Matt has $100 and 75 Euros.
Which is worth more – the dollars or the Euros?

7 Work out the following if $a = 1.2$, $b = 6.8$ and $c = 7.1$.
   a $ab + c$ b $3(ab + ac + bc)$

8 Work out:

a $\sqrt{(0.8^2 + 1.5^2)}$ b $\sqrt{(5.2^2 - 2^2)}$

9 Work out:

a $6.5^3 \times 2 - 2 \times 8.1$

b $2.66^3 - 3 \div 0.15 + 6.4$

1.2 Adding and subtracting fractions with a calculator

HOMEWORK 1B

1 Use your calculator to work out the following.

Try to key in the calculation as one continuous set, without writing down any intermediate values.

Give your answers as mixed fractions.

a $5\frac{1}{4} + 7\frac{3}{5}$ b $8\frac{2}{3} + 1\frac{4}{9}$ c $6\frac{3}{4} + 2\frac{7}{10}$

d $9\frac{1}{8} + 3\frac{7}{25}$ e $7\frac{9}{20} + 3\frac{5}{16}$ f $8\frac{3}{8} + 1\frac{3}{16} + 2\frac{3}{4}$

g $6\frac{17}{20} - 5\frac{5}{12}$ h $2\frac{5}{8} - 1\frac{7}{24}$ i $3\frac{9}{32} - 1\frac{1}{12}$

j $4\frac{3}{5} + 5\frac{7}{16} - 8\frac{1}{3}$ k $1\frac{9}{24} + 1\frac{5}{18} - 1\frac{1}{10}$ l $5\frac{1}{4} + 2\frac{7}{9} - 6\frac{5}{13}$

AU 2 A tank of water is empty. Two-thirds of a full tank is poured in. One-quarter of a full tank is poured out. One-twelfth of a full tank is poured in.

What fraction of the tank is now full of water?

3 Look at this road sign.

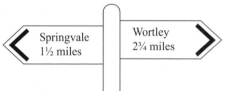

Springvale
1½ miles

Wortley
2¾ miles

a What is the distance between Springvale and Wortley using these roads?

b How much further is it to Wortley than to Springvale?

AU 4 Here is a calculation:

$\frac{1}{4} \times \frac{2}{3}$

Imagine that you are trying to explain to someone how to do this using a calculator.

Write down what you would say.

PS 5 A class has the same numbers of boys and girls.

Three girls leave and three boys join the class.

The fraction of the class who are girls is now $\frac{3}{8}$.

How many are in the class?

AU 6 a Use your calculator to work out $\frac{19}{23} - \frac{21}{25}$.

b Explain how your answer tells you that $\frac{19}{23}$ is less than $\frac{21}{25}$.

AU 7 a Work out $\frac{10}{27} - \frac{3}{11}$ on your calculator.

b Work out $\frac{10}{27} - \frac{7}{16}$ on your calculator.

c Explain why your answers to parts **a** and **b** show that $\frac{10}{27}$ is a fraction in between $\frac{3}{11}$ and $\frac{7}{16}$.

8 To work out the circumference of a circle, the following formula is used.

$C = \pi d$

where d is the diameter.

Work out the circumference of a circle when the diameter is 9 cm.

PS 9 A shape is rotated 30° clockwise and then 90° anticlockwise.

What fraction of a turn is needed to return it to its original position?

Give both possible answers.

1.3 Multiplying and dividing fractions with a calculator

HOMEWORK 1C

1 Use your calculator to work out the following.

Try to key in the calculation as one continuous set, without writing down any intermediate values.

Give your answers as fractions.

a $\frac{1}{4} \times \frac{3}{5}$ **b** $\frac{2}{3} \times \frac{4}{9}$ **c** $\frac{3}{4} \times \frac{7}{10}$

d $\frac{1}{8} \times \frac{7}{25}$ **e** $\frac{9}{20} \times \frac{5}{16}$ **f** $\frac{3}{8} \times \frac{3}{16} \times \frac{3}{4}$

g $\frac{17}{20} \div \frac{5}{12}$ **h** $\frac{5}{8} \div \frac{7}{24}$ **i** $\frac{9}{32} \div \frac{1}{12}$

j $\frac{3}{5} \times \frac{7}{16} \div \frac{1}{3}$ **k** $\frac{9}{24} \times \frac{5}{18} \div \frac{1}{10}$ **l** $\frac{1}{4} \times \frac{7}{9} \div \frac{5}{13}$

2 The formula for the area of a rectangle is:

Area = length × width

Use this formula to work the area of a rectangle of length $\frac{3}{4}$ metres and width $\frac{1}{3}$ metres.

3 Bricks are $\frac{1}{6}$ metre long.

How many bricks placed end to end would be needed to make a line two metres long?

AU 4 **a** Use your calculator to work out $\frac{2}{3} \times \frac{7}{11}$ **b** Write down the answer to $\frac{2}{11} \times \frac{7}{3}$

AU 5 **a** Use your calculator to work out $\frac{3}{4} \div \frac{7}{12}$ **b** Use your calculator to work out $\frac{3}{4} \times \frac{12}{7}$

        c Use your calculator to work out $\frac{2}{9} \div \frac{2}{3}$ **d** Write down the answer to $\frac{2}{9} \times \frac{3}{2}$

6 Use your calculator to work out the following questions. Try to key in the calculation as one continuous set, without writing down any intermediate values.

Give your answers as mixed fractions.

a $3\frac{1}{4} \times 2\frac{3}{5}$ **b** $6\frac{2}{3} \times 1\frac{4}{9}$ **c** $7\frac{3}{4} \times 2\frac{7}{10}$

d $5\frac{1}{8} \times 2\frac{7}{25}$ **e** $6\frac{9}{20} \times 4\frac{5}{16}$ **f** $1\frac{3}{8} \times 1\frac{3}{16} \times 1\frac{3}{4}$

g $4\frac{17}{20} \div 2\frac{5}{12}$ **h** $1\frac{5}{8} \div 1\frac{7}{24}$ **i** $2\frac{9}{32} \div 1\frac{1}{12}$

j $3\frac{3}{5} \times 2\frac{7}{16} \div 1\frac{1}{3}$ **k** $2\frac{9}{24} \times 3\frac{5}{18} \div 1\frac{1}{10}$ **l** $4\frac{1}{4} \times 3\frac{7}{9} \div 2\frac{5}{13}$

7 The formula for the area of a rectangle is:

Area = length × width

Use this formula to work the area of a rectangle of length $4\frac{3}{4}$ metres and width $2\frac{1}{3}$ metres.

8 The volume of a sphere is $19\frac{2}{5}$ cm³. It is cut into four equal pieces.

Work out the volume of one of the pieces.

9 The formula for average speed is:

Average speed = Distance ÷ time taken

Work out the average speed of a car which travels $6\frac{3}{4}$ miles in a $\frac{1}{4}$ of an hour.

10 Given that 1 gallon = $4\frac{1}{2}$ litres

Grace puts 40 litres of fuel in her car.

How many gallons is this?

Give your answer to the nearest gallon.

 11 Ropes come in $12\frac{1}{2}$ metre lengths. Jack wants to cut pieces of rope that are each $\frac{3}{8}$ of a metre long.

He needs 100 pieces.

How many ropes will he need?

Functional Maths Activity

Calculating a gas bill

The following information is written on the back of Mr Fermat's gas bill.

Reading on 19th Aug   05979
Reading on 19th Nov   06229

= 250 metric units used over 93 days

Gas units converted   = 2785.52 kWh used over 93 days

First 683.00 kWh × 6.683p   £45.64
Next 2102.52 kWh × 3.292p   £69.21

Total cost of gas used   £114.85

Gas units are converted to kilowatt hours (kWh) using the following formula:

Metric units used	calorific value correction	volume	to convert to kWh	gas used in kWh
250	× 39.2236	× 1.02264	÷ 3.6	= 2785.52

Mr Fermat is having trouble understanding this and has asked for your help. Can you answer these questions for him?

1   Where does the figure of 250 units come from?
2   What does kWh stand for?
3   There are two different prices for gas. The first 683.00 kWh used are charged at a higher rate than the 2102.52 kWh used after that. What are the two different prices for each kWh?
4   Can you check that the formula at the bottom has been worked out correctly: does 250 metric units convert to 2785.52 kWh of gas used?
5   What is the average cost per day of the gas Mr Fermat has used?
6   In fact, the reading taken on 19th November was an estimate because Mr Fermat was out when the meter reader called. When Fermat looks at the meter, the reading is only 06203. By how much has he been overcharged?
7   Mr Fermat informs the gas company of the correct reading and they recalculate his bill. What is the percentage reduction in his bill?

2.1 Volume of a pyramid

HOMEWORK 2A

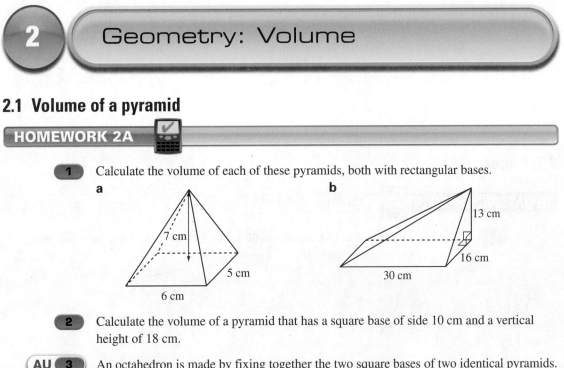

1 Calculate the volume of each of these pyramids, both with rectangular bases.

a

7 cm

5 cm

6 cm

b

13 cm

16 cm

30 cm

2 Calculate the volume of a pyramid that has a square base of side 10 cm and a vertical height of 18 cm.

AU 3 An octahedron is made by fixing together the two square bases of two identical pyramids. Each pyramid is 9 cm high and has a base with each side 7 cm. Calculate the volume of the octahedron.

4 The Khufu pyramid in Egypt was originally 146 m tall.
Each side of the square base was 230 m long.
It was built from limestone blocks with a density of about 2.7 tonnes per cubic metre.
It probably took 20 years to complete.
Estimate the total weight of the blocks used to build the pyramid.

5 Calculate the volume of this shape.

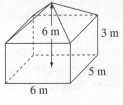

6 m   3 m

5 m

6 m

6 Calculate the height *h* of a rectangular-based pyramid with a length of 14 cm, a width of 10 cm and a volume of 140 cm³.

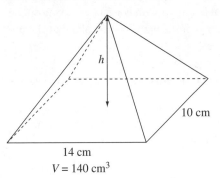

h

10 cm

14 cm

$V = 140 \text{ cm}^3$

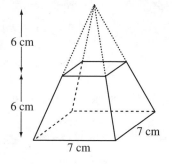

7 The pyramid in the diagram has its top cut off as shown. The shape which is left is called a frustum. Calculate the volume of the frustum.

6 cm

6 cm

7 cm

7 cm

2.2 Cones

HOMEWORK 2B

1 For each cone, calculate **i** its volume and **ii** its total surface area. (The units are cm.)

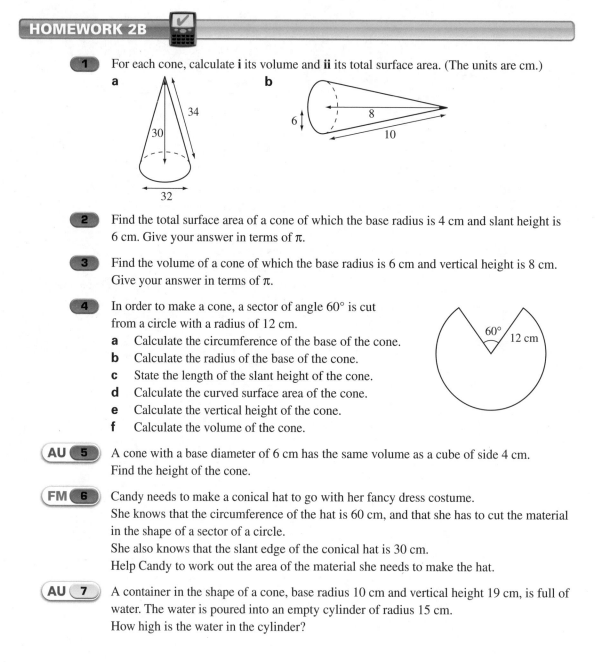

a

34

30

32

b

6

8

10

2 Find the total surface area of a cone of which the base radius is 4 cm and slant height is 6 cm. Give your answer in terms of π.

3 Find the volume of a cone of which the base radius is 6 cm and vertical height is 8 cm. Give your answer in terms of π.

4 In order to make a cone, a sector of angle 60° is cut from a circle with a radius of 12 cm.
  a Calculate the circumference of the base of the cone.
  b Calculate the radius of the base of the cone.
  c State the length of the slant height of the cone.
  d Calculate the curved surface area of the cone.
  e Calculate the vertical height of the cone.
  f Calculate the volume of the cone.

60°

12 cm

AU 5 A cone with a base diameter of 6 cm has the same volume as a cube of side 4 cm. Find the height of the cone.

FM 6 Candy needs to make a conical hat to go with her fancy dress costume.
She knows that the circumference of the hat is 60 cm, and that she has to cut the material in the shape of a sector of a circle.
She also knows that the slant edge of the conical hat is 30 cm.
Help Candy to work out the area of the material she needs to make the hat.

AU 7 A container in the shape of a cone, base radius 10 cm and vertical height 19 cm, is full of water. The water is poured into an empty cylinder of radius 15 cm.
How high is the water in the cylinder?

PS 8 The diagram shows a paper cone. The diameter of the base is 4.8 cm and the slant height is 4 cm. The cone is cut along the line AV and opened out flat, as shown below.

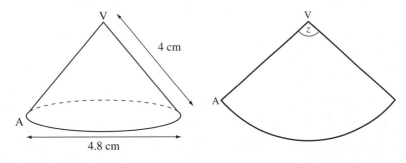

Calculate the size of angle *z*.

2.3 Spheres

HOMEWORK 2C

1 Calculate the volume of spheres with the following measurements.
Give your answers in terms of π.
 a Radius 3 cm   **b** Diameter 30 cm

2 Calculate the surface area of spheres with the following measurements.
Give your answers in terms of π.
 a Radius 4 cm   **b** Diameter 10 cm

3 Calculate the volume and the surface area of a sphere with a diameter of 30 cm.
Give your answers to a suitable degree of accuracy.

4 Calculate, correct to one decimal place, the radius of a sphere:
 a with a surface area of 200 cm^2 **b** with a volume of 200 cm^3

5 The volume of a sphere is 50 m^3. Find its diameter.

6 What is the volume of a sphere with a surface area of 400 cm^2?

AU 7 A cube of metal of side 5 cm has a hemispherical hole of diameter 4 cm cut into it.
What is the volume of the resulting shape?

FM 8 A roller skate manufacturing company needs to make 4 mm-diameter steel ball bearings.
How many ball bearings can the company make from one cubic metre of steel?

9 A spinning top, which consists of a cone of base radius 6 cm,
slant height 10 cm and a hemisphere of radius 6 cm, is
illustrated on the right. Give your answers in terms of π.
 a Calculate the volume of the spinning top.
 b Calculate the total surface area of the spinning top.

Functional Maths Activity

Golf balls

Until 1990 a golf ball had to be no less than 1.62 inches in diameter. Then the regulations were changed, and now a golf ball has to be no less than 1.68 inches in diameter. Sizes are still given in inches.

1   How has the volume of a minimum size golf ball changed?

2   What is the percentage increase in the volume of the minimum size of golf ball?

3   Why would your answer to question 2 be of interest to a golf ball manufacturer?

Dimples on the surface of a ball were first introduced in 1908. They improve the flight of the ball. they must be arranged 'as symmetrically as possible'. Most balls have been between 250 and 450 dimples.

4   If an old (1.62 inch) ball had 400 dimples, how many dimples of the same size could you fit onto a new (1.68 inch) ball?

3.1 Trigonometric ratios

HOMEWORK 3A

In these questions, give any answers involving angles to the nearest degree.

1 Find these values, rounding off your answers to 3 significant figures.
  a sin 52°   **b** sin 46°   **c** sin 76.3°   **d** sin 90°

2 Find these values, rounding off your answers to 3 significant figures.
  a cos 52°   **b** cos 46°   **c** cos 76.3°   **d** cos 90°

3 **a** Calculate $(\sin 52°)^2 + (\cos 52°)^2$ **b** Calculate $(\sin 46°)^2 + (\cos 46°)^2$
  c Calculate $(\sin 76.3°)^2 + (\cos 76.3°)^2$ **d** Calculate $(\sin 90°)^2 + (\cos 90°)^2$
  e What do you notice about your answers?

4 Use your calculator to work out the value of:
  a tan 52°   **b** tan 46°   **c** tan 76.3°   **d** tan 0°

5 Use your calculator to work out the value of:
  a sin 52° ÷ cos 52°   **b** sin 46° ÷ cos 46°   **c** sin 76.3° ÷ cos 76.3°
  d sin 0° ÷ cos 0°
  e What connects your answers with the answers to Question **4**?

6 Use your calculator to work out the value of:
  a 6 sin 55°   **b** 7 cos 45°   **c** 13 sin 67°   **d** 20 tan 38°

7 Use your calculator to work out the value of:
  a $\dfrac{6}{\sin 55°}$ **b** $\dfrac{7}{\cos 45°}$ **c** $\dfrac{13}{\sin 67°}$ **d** $\dfrac{20}{\tan 38°}$

8 Using the following triangle, calculate sin, cos, and tan for the angle marked x. Leave your answers as fractions.

AU 9 You are given that $\sin x = \dfrac{5}{\sqrt{34}}$. Work out the value of $\tan x$.

3.2 Calculating angles

HOMEWORK 3B

Use your calculator to find the answers to the following to one decimal place.

1 What angles have sines of:
  a 0.4   **b** 0.707   **c** 0.879   **d** 0.666666666666666...

2 What angles have cosines of:
  a 0.4   **b** 0.707   **c** 0.879   **d** 0.333333333333333...

FM Functional Maths **AU** (AO2) Assessing Understanding **PS** (AO3) Problem Solving

3 What angles have the following tangents?

    a   0.4   **b**   1.24   **c**   0.875   **d**   2.625

4 What angles have the following sines?

    a $3 \div 8$ **b** $1 \div 3$ **c** $3 \div 10$ **d** $5 \div 8$

5 What angles have the following cosines?

    a $3 \div 8$ **b** $1 \div 3$ **c** $3 \div 10$ **d** $5 \div 8$

6 What angles have the following tangents?

    a $3 \div 8$ **b** $3 \div 2$ **c** $5 \div 7$ **d** $19 \div 5$

7 If sin 54° = 0.809 to 3 decimal places, what angle has a cosine of 0.809?

3.3 Using the sine and cosine functions

HOMEWORK 3C

1 Find the value marked *x* in each of these diagrams.

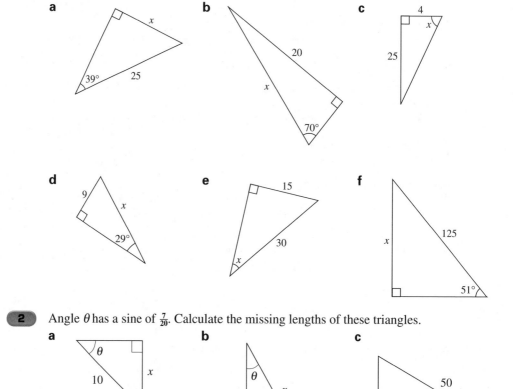

2 Angle θ has a sine of $\frac{7}{20}$. Calculate the missing lengths of these triangles.

AU **3** Caxton is due north of Ashville and due west of Peaton. A pilot flies directly from Ashville to Peaton, a distance of 15 km, on a bearing of 050°.

a Calculate the direct distance from Caxton to Peaton.

b Work out the bearing of Ashville from Peaton.

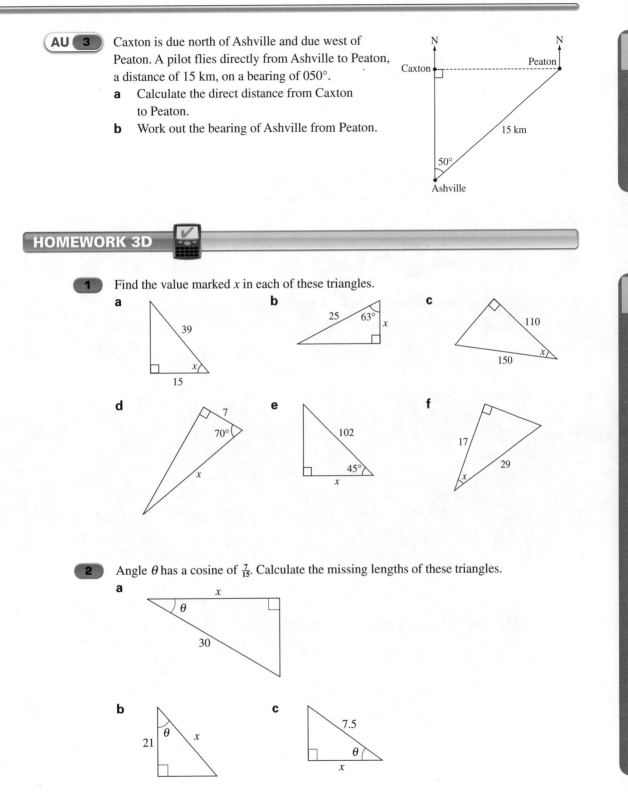

HOMEWORK 3D

1 Find the value marked x in each of these triangles.

a

b

c

d

e

f

2 Angle θ has a cosine of $\frac{7}{15}$. Calculate the missing lengths of these triangles.

a

b

c

B

AU 3 The diagram shows the positions of three
telephone masts A, B and C.
Mast C is 6 kilometres due east of Mast B.
Mast A is due north of Mast B, and 9 km from
Mast C.

a Calculate the distance of A from B.
Give your answer in kilometres, correct
to 3 significant figures.

b Calculate the size of the angle marked x.
Give your angle correct to one decimal place.

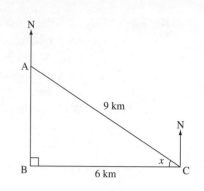

3.4 Using the tangent function

HOMEWORK 3E

1 Find the value marked x in each of these triangles.

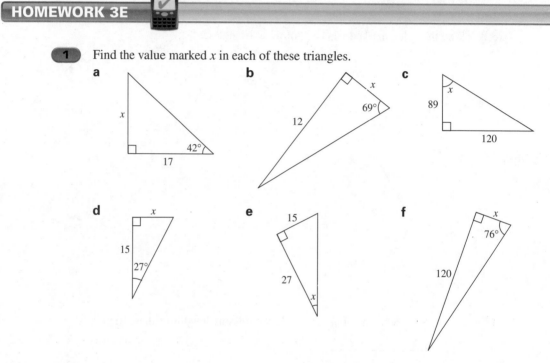

a

b

c

d

e

f

2 Angle θ has a tangent of $\frac{2}{3}$. Calculate the missing lengths of these triangles.

a

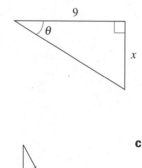

b

c

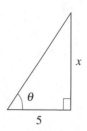

AU **3** The sensor for a security light is fixed to a house wall 2.25 m above the ground. It can detect movement on the ground up to 15 m away from the house. B is the furthest point where the sensor, A, can detect movement.
Calculate the size of angle *x*.

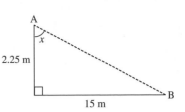

3.5 Which ratio to use

HOMEWORK 3F

1 Find the angle or length marked *x* in each of these triangles.

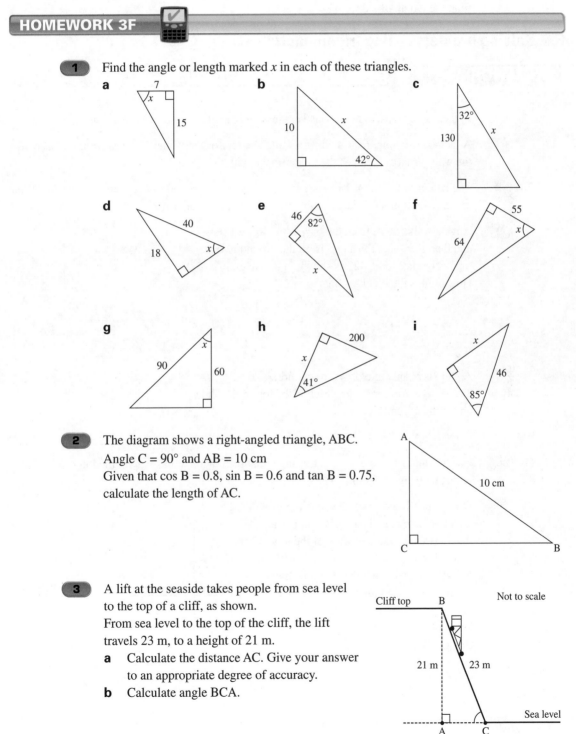

a

b

c

d

e

f

g

h

i

2 The diagram shows a right-angled triangle, ABC.
Angle C = 90° and AB = 10 cm
Given that cos B = 0.8, sin B = 0.6 and tan B = 0.75, calculate the length of AC.

3 A lift at the seaside takes people from sea level to the top of a cliff, as shown.
From sea level to the top of the cliff, the lift travels 23 m, to a height of 21 m.
a Calculate the distance AC. Give your answer to an appropriate degree of accuracy.
b Calculate angle BCA.

4 Look at this triangle.

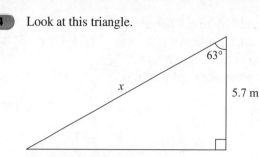

Find the length of side x.

3.6 Solving problems using trigonometry 1

HOMEWORK 3G

In these questions, give any answers involving angles to the nearest degree.

1 A ladder, 8 m long, rests against a wall. The foot of the ladder is 2.7 m from the base of the wall. What angle does the ladder make with the ground?

FM 2 The ladder in Question **1** has a 'safe angle' with the ground of between 70° and 80°. What are the safe limits for the distance of the foot of the ladder from the wall?

3 Angela walks 60 m from the base of a block of flats and then measures the angle from the ground to the top of the flats to be 42° as shown in the diagram. How high is the block of flats?

4 A slide is at an angle of 46° to the horizontal. The slide is 7 m long. How high is the top of the slide above the ground?

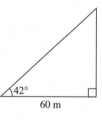

5 Use trigonometry to calculate the angle that the diagonal makes with the long side of a rectangle 9 cm by 5 cm.

FM 6 Drumsbury Town Council wants to put up a flag pole outside the town hall. The diagram shows the end view of the town hall building.
Regulations state that the flag pole must not be more than half the height of the building.
What is the maximum height that the flag pole can be?

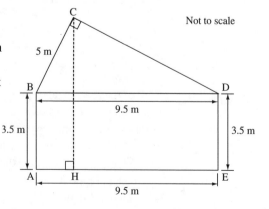

Not to scale

AU 7 A road rises steadily at an angle of 6°. A lorry travels 300 m along the road. What is the increase in height?

8 A swing at rest is 50 cm above the ground and 2 m below the point of suspension. When a child is on the swing, the angle with the vertical can be as large as 35°.

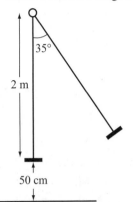

The child's father thinks that she might then be dangerously far from the ground. Can you tell him exactly how far his daughter will be above the ground?

HOMEWORK 3H

In these questions, give any answers involving angles to the nearest degree.

1 Eric sees an aircraft in the sky. The aircraft is at a horizontal distance of 15 km from Eric. The angle of elevation is 42°. How high is the aircraft?

2 A man standing 100 m from the base of a block of flats, looks at the top of the block and notices that the angle of elevation is 49°. How high is the block of flats?

3 A man stands 15 m from a tree. The angle of elevation of the top of the tree from his eye is 25°. If his eye is 1.5 m above the ground, how tall is the tree?

4 A bird, sitting at the very top of the tree in Question **3**, sees a worm next to the foot of the man. What is the angle of depression from the bird's eye to the worm?

5 I walk 200 m away from a chimney that is 120 m high. What is the angle of elevation from my eye to the top of the chimney? (Ignore the height of my eye above the ground.)

6 If you are now told that the height of my eye in Question **5** is 1.8 m above ground, how much different is the angle of elevation?

AU 7 Boat B is moored 50 m from the foot of a vertical cliff. The angle of depression of the boat from the top of the cliff is 52°.
   a Calculate the height of the cliff.
   b The boat is released from its mooring and it drifts 350 m further away from the cliff. Calculate the angle of elevation of the top of the cliff from the boat.

Cliff

B————

50 m

8 A boat is 450 m from the base of a cliff. The angle of elevation of the top of the cliff is 8°. How high is the cliff?

FM **9**   To find the height of a tree, Sacha tries to measure the angle of elevation of the top from a point 40 m away.

He finds it difficult to measure the angle accurately, but thinks it is between 30° and 35°. What can you tell him about the height of the tree?

3.7 Solving problems using trigonometry 2

HOMEWORK 3I

1   A ship sails for 85 km on a bearing of 067°.

    a   How far east has it travelled?

    b   How far north has the ship sailed?

2   Rotherham is 11 miles south of Barnsley and 2 miles west of Barnsley.

    What is the bearing of:

    a   Barnsley from Rotherham   **b**   Rotherham from Barnsley?

3   A plane sets off from airport A and flies due east for 100 km, then turns to fly due south for 80 km before landing at an airport B. What is the bearing of airport B from airport A?

FM **4**   Mountain A is due east of a walker. Mountain B is due south of the walker.

    The guidebook says that mountain A is 5 km from mountain B, on a bearing of 038°.

    How far is the walker from mountain B?

5   The diagram shows the relative distances and bearings of three ships A, B and C.

    a   How far north of A is B?

         (Distance x on diagram.)

    b   How far north of B is C?

         (Distance y on diagram.)

    c   How far west of A is C?

         (Distance z on diagram.)

    d   What is the bearing of A from C?

         (Angle w on diagram.)

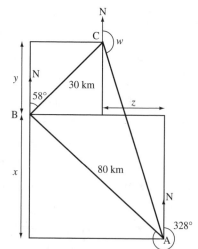

AU **6**   An aeroplane is flying from Leeds (L) to London Heathrow (H). It flies 150 miles on a bearing 136° to point A. It then turns through 90° and flies the final 80 miles to H.

    a   **i**   Show clearly why the angle marked x is equal to 46°.

         ii   Give the bearing of H from A.

    b   Use Pythagoras' theorem to calculate the distance LH.

    c   **i**   Calculate the size of the angle marked y.

         ii   Work out the bearing of L from H.

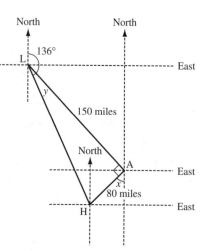

AU 7 A plane flies 200 km on a bearing of 124° and then 150 km on a bearing of 053°.
How far east from its starting point has it travelled?

FM 8 Large boats are supposed to stay at least 300 m from the shore near a particular beach.
Don notices a large boat that is due north from where he is sitting on the beach.
He walks 100 m to the east and uses a compass to find that the bearing of the boat is 340°.
Is the boat breaking the rules?

HOMEWORK 3J

1 Find the side or angle marked x.

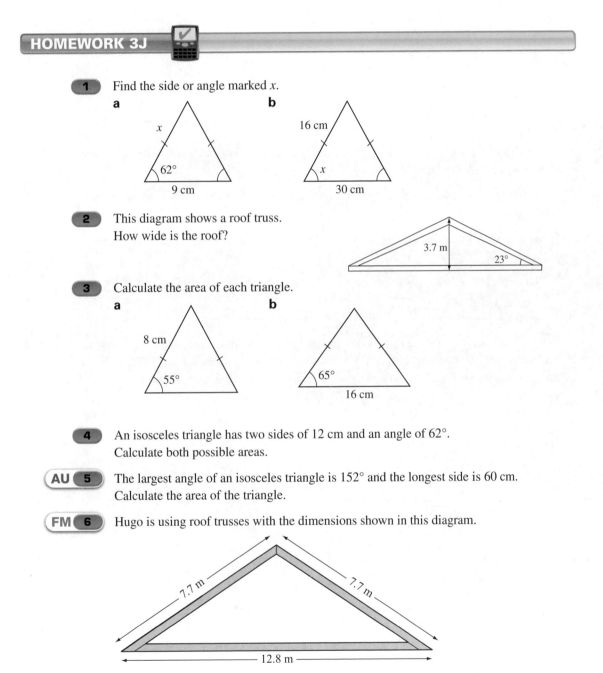

a

x

62°

9 cm

b

16 cm

x

30 cm

2 This diagram shows a roof truss.
How wide is the roof?

3.7 m

23°

3 Calculate the area of each triangle.

a

8 cm

55°

b

65°

16 cm

4 An isosceles triangle has two sides of 12 cm and an angle of 62°.
Calculate both possible areas.

AU 5 The largest angle of an isosceles triangle is 152° and the longest side is 60 cm.
Calculate the area of the triangle.

FM 6 Hugo is using roof trusses with the dimensions shown in this diagram.

7.7 m   7.7 m

12.8 m

What is the angle of slope of the roof?

Problem-solving Activity

Investigating spirals

An interesting spiral shape can be constructed with right-angled triangles, as shown in the diagram below.

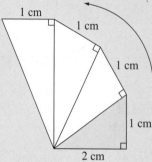

To make this shape, start with a right-angled triangle of base 2 cm and height 1 cm (the shaded triangle at the bottom of the diagram).

The second right-angled triangle is built on top of the hypotenuse of the first triangle, and has a height of 1 cm.

The shape grows by continuing to put right-angled triangles (each with a height of 1 cm) on top, as shown in the diagram.

1 Draw the shape as far as you can go.
2 Measure the hypotenuse of the last triangle to be drawn.
3 Now calculate what that length should have been.

Now look at the following shape, which is built up from a first right-angled triangle which has a hypotenuse of 5 cm and height of 1 cm. This time, each triangle added underneath is built under the base of the previous one, with the previous base length becoming the new hypotenuse.

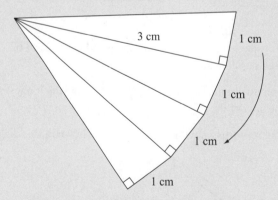

4 Continue drawing the shape as far as you think you can.
5 Measure the length of the smallest base that you end up with.
6 Calculate the length of the final base.

Geometry: Properties of circles

4.1 Circle theorems

HOMEWORK 4A

1 Find the value of *x* in each of these circles with centre O.

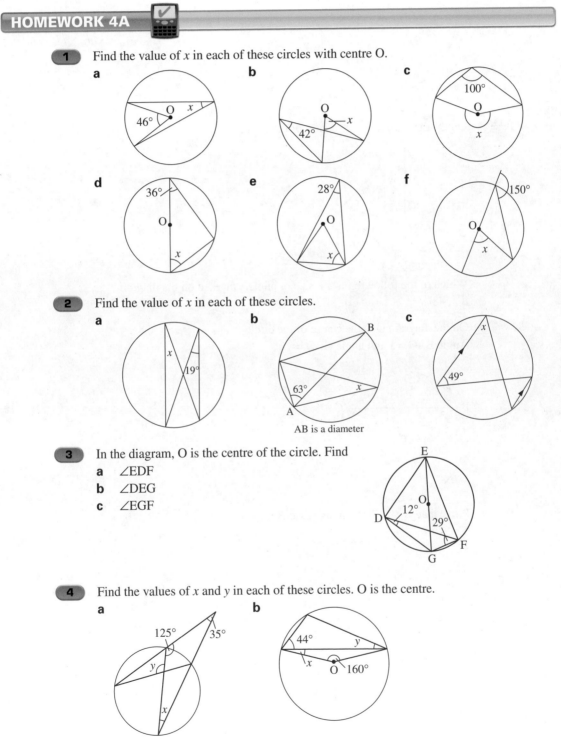

a

46°   O   *x*

b

O   *x*   42°

c

100°   O   *x*

d

36°   O   *x*

e

28°   O   *x*

f

150°   O   *x*

2 Find the value of *x* in each of these circles.

a

x   19°

b

B   63°   *x*   A

AB is a diameter

c

x   49°

3 In the diagram, O is the centre of the circle. Find

 a ∠EDF

 b ∠DEG

 c ∠EGF

E   O   12°   D   29°   F   G

4 Find the values of *x* and *y* in each of these circles. O is the centre.

a

125°   35°   *y*   *x*

b

44°   *y*   *x*   O   160°

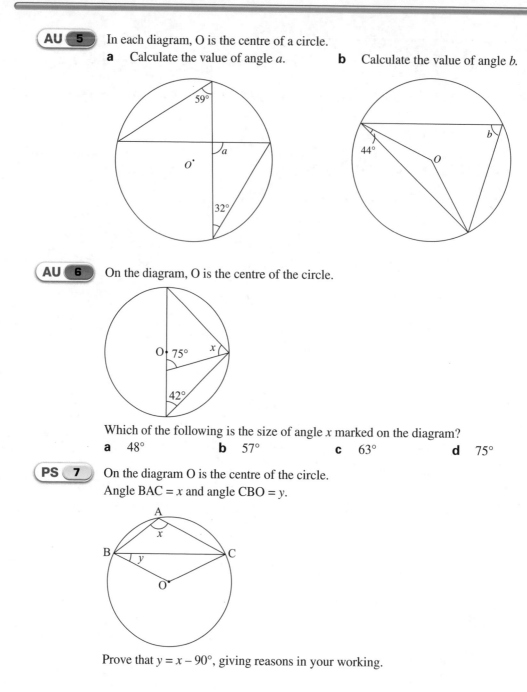

AU **5** In each diagram, O is the centre of a circle.
   a Calculate the value of angle *a*.   **b** Calculate the value of angle *b*.

AU **6** On the diagram, O is the centre of the circle.

Which of the following is the size of angle *x* marked on the diagram?
   a 48°   **b** 57°   **c** 63°   **d** 75°

PS **7** On the diagram O is the centre of the circle.
Angle BAC = *x* and angle CBO = *y*.

Prove that $y = x - 90°$, giving reasons in your working.

4.2 Cyclic quadrilaterals

HOMEWORK 4B

1 Find the size of the lettered angles in each of these circles.

a

80°
112°
a   b

b

d   e
f   82°

c

e   85°
d   111°

d

m
n
118°
38°

2 Find the values of x and y in each of these circles.

a

x
91°

b

B
33°
65°
x
A

c

x
49°
y

3 Find the values of x and y in each of these circles, centre O.

a

x
O
y
52°

b

60°
O
y
x

c

y
95°
105°
x

PS 4 ABCD is a cyclic quadrilateral.

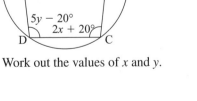

A   B
3y + 5°   2x − 5°
5y − 20°
2x + 20°
D   C

Work out the values of x and y.

AU 5 On the diagram, O is the centre of the circle.
Explain why the angle BOD is 128°.
Give reasons for your answer.

6 ABCD are points on a circle. AB is parallel to CD.
Prove that $x = y$

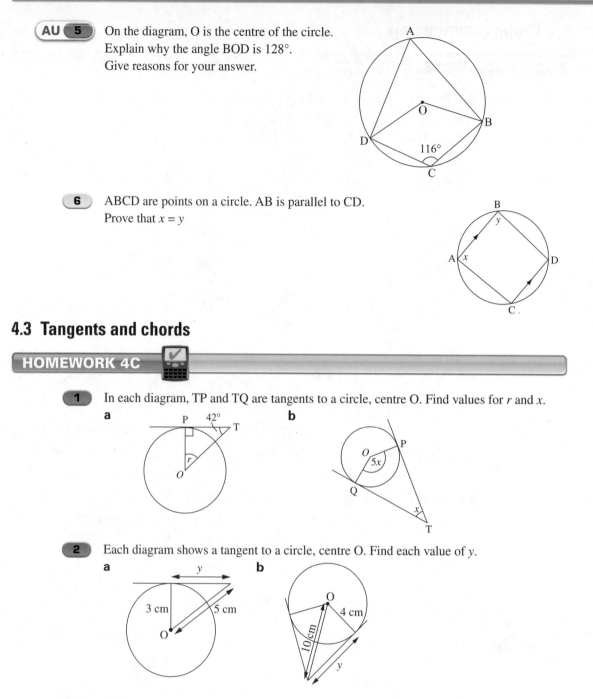

4.3 Tangents and chords

HOMEWORK 4C

1 In each diagram, TP and TQ are tangents to a circle, centre O. Find values for r and x.

a   **b**

2 Each diagram shows a tangent to a circle, centre O. Find each value of y.

a   **b**

3 Each diagram shows a tangent to a circle, centre O. Find x and y in each case.

a   **b**

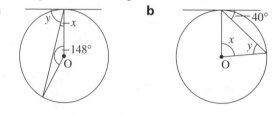

4 In each of the diagrams, TP and TQ are tangents to the circle, centre O. Find each value of *x*.

a

b

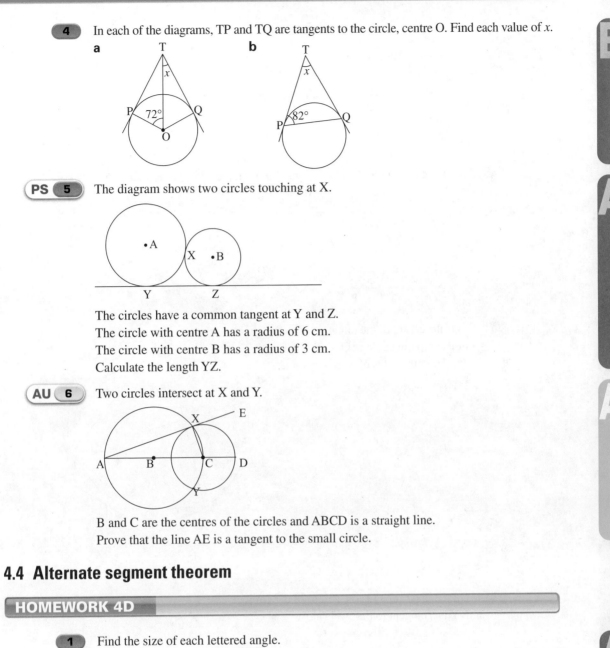

PS 5 The diagram shows two circles touching at X.

The circles have a common tangent at Y and Z.
The circle with centre A has a radius of 6 cm.
The circle with centre B has a radius of 3 cm.
Calculate the length YZ.

AU 6 Two circles intersect at X and Y.

B and C are the centres of the circles and ABCD is a straight line.
Prove that the line AE is a tangent to the small circle.

4.4 Alternate segment theorem

HOMEWORK 4D

1 Find the size of each lettered angle.

a

b

2 In each diagram, find the value of x.

a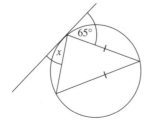

b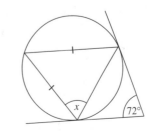

3 In each diagram, find the value of *x* and *y*.

a   **b**

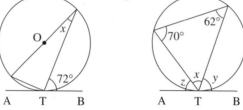

4 ATB is a tangent to the circle, centre O. Find the values of *x*, *y* and *z* in each case.

a   **b**

5 O is the centre of the circle. PQT is the
tangent to the circle at Q.
Work out the sizes of angles *x*, *y* and *z*.

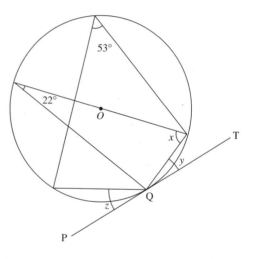

6 On the diagram, O is the centre of the circle.
XY is a tangent to the circle at A.

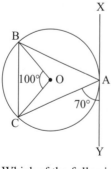

Which of the following is the size of angle OBA?

a   25°   **b**   30°   **c**   35°   **d**   40°

7 AB is a tangent to the circle at X.
YZC is a straight line.

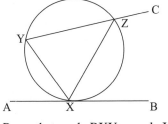

Prove that angle BXY = angle XZC.

Functional Maths Activity

Security cameras

A museum has a circular room which is used to show some of its valuable exhibits.

The room needs to be completely covered by security cameras placed around the wall of the room.

The curator of the museum wants to use as few cameras as possible to keep the cost low and to make the supervision easier.

The curator is thinking of using security cameras with an angle of view of 60°.

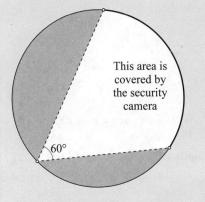

This area is covered by the security camera

60°

1 How many cameras will the curator need for the room?
2 How should the cameras be arranged?

5 Number: Powers, standard form and surds

5.1 Powers (indices)

HOMEWORK 5A

1 Write these expressions using power notation. Do not work them out yet.
- **a** $5 \times 5 \times 5 \times 5$
- **b** $7 \times 7 \times 7 \times 7 \times 7$
- **c** $19 \times 19 \times 19$
- **d** $4 \times 4 \times 4 \times 4 \times 4$
- **e** $1 \times 1 \times 1 \times 1 \times 1 \times 1 \times 1$
- **f** $8 \times 8 \times 8 \times 8 \times 8$
- **g** 6
- **h** $11 \times 11 \times 11 \times 11 \times 11 \times 11$
- **i** $0.9 \times 0.9 \times 0.9 \times 0.9$
- **j** $999 \times 999 \times 999$

2 Write these power terms out in full. Do not work them out yet.
- **a** 4^5
- **b** 8^4
- **c** 5^3
- **d** 9^6
- **e** 1^{11}
- **f** 7^3
- **g** 5.2^3
- **h** 7.5^3
- **i** 7.7^4
- **j** $10\,000^3$

3 Using the power key on your calculator (or another method), work out the values of the power terms in Question **1**.

4 Using the power key on your calculator (or another method), work out the values of the power terms in Question **2**.

FM 5 A box is a cube.
The width of the box is 1 metre.
To work out the volume of a box, use the formula:
Volume = (width)^3
Some dolls are manufactured and packed into the box. The volume of each doll is 0.03 m^3.
Work out the amount of space in the box when 24 dolls are packed into the box.

AU 6 Write each number as a power of a different number.
The first one has been done for you.
- **a** $27 = 3^3$
- **b** $16 =$
- **c** $125 =$
- **d** $64 =$

7 Without using a calculator, work out the values of these power terms.
- **a** 7^0
- **b** 9^1
- **c** 17^0
- **d** 1^{91}
- **e** 10^5

8 Using your calculator, or otherwise, work out the values of these power terms.
- **a** $(-2)^3$
- **b** $(-1)^{11}$
- **c** $(-3)^4$
- **d** $(-5)^3$
- **e** $(-10)^6$

9 Without using a calculator, write down the answers to these.
- **a** $(-4)^2$
- **b** $(-5)^3$
- **c** $(-3)^4$
- **d** $(-2)^5$
- **e** $(-1)^6$

PS 10 $5^5 = 3125$ $5^6 = 15\,625$ $5^7 = 78\,125$ $5^8 = 390\,625$
Write down the last three digits of each of the following powers of 5.
- **a** 5^{99}
- **b** 5^{100}

HOMEWORK 5B

AU 1 You are given that $6^4 = 1296$.
Write down the value of 6^{-4}.

2 Write down each of these in fraction form.
  a 5^{-2} **b** 4^{-1} **c** 10^{-3} **d** 3^{-3} **e** x^{-2} **f** $5t^{-1}$

3 Write down each of these in negative index form.
  a $\dfrac{1}{2^4}$ **b** $\dfrac{1}{7}$ **c** $\dfrac{1}{x^2}$

4 Change each of the following expressions into an index form of the type shown.
  a All of the form 2^n **i** 32   **ii** $\frac{1}{4}$
  b All of the form 10^n **i** 10 000   **ii** $\frac{1}{100}$
  c All of the form 5^n **i** 625   **ii** $\frac{1}{125}$

5 Find the value of each of the following, where the letters have the given values.
  a Where $x = 3$ **i** x^2 **ii** $4x^{-1}$
  b Where $t = 5$ **i** t^{-2} **ii** $5t^{-4}$
  c Where $m = 2$ **i** m^{-3} **ii** $4m^{-2}$

6 $a = 3$ and $b = 2$. Calculate the value of
  a $3a^{-1} + 2b^{-2}$, giving your answer as a fraction in its simplest form.
  b $a^{-2} + b^{-3}$, giving your answer as a fraction in its simplest form.

PS 7 a and b are integers.
$2^a + 3^b = 41$
Work out the values of a and b.

AU 8 c and d are integers. c is even.
Decide whether $c^2 + d^3$ is even or odd, or whether it could be either. Give examples to show how you decided.

PS AU 9 Put these in order from smallest to largest:
x^0 x^{-1} x^1
  a when x is greater than 1
  b when x is between 0 and 1
  c when x is between -1 and 0

HOMEWORK 5C

1 Write these as single powers of 7.
  a $7^3 \times 7^2$ **b** $7^3 \times 7^6$ **c** $7^4 \times 7^3$ **d** 7×7^5 **e** $7^5 \times 7^9$ **f** 7×7^7

2 Write these as single powers of 5.
  a $5^6 \div 5^2$ **b** $5^8 \div 5^2$ **c** $5^4 \div 5^3$ **d** $5^5 \div 5^5$ **e** $5^6 \div 5^4$

3 Simplify these and write them as single powers of a.
  a $a^2 \times a$ **b** $a^3 \times a^2$ **c** $a^4 \times a^3$ **d** $a^6 \div a^2$ **e** $a^3 \div a$ **f** $a^5 \div a^4$

AU 4   **a** $a^x \times a^y = a^6$

Write down a possible pair of values of x and y.

b $a^x \div a^y = a^6$

Write down a possible pair of values of x and y.

5   Simplify these expressions.

a $3a^4 \times 5a^2$ **b** $3a^4 \times 7a$ **c** $5a^4 \times 6a^2$ **d** $3a^2 \times 4a^7$

e $5a^4 \times 5a^2 \times 5a^2$

6   Simplify these expressions.

a $8a^5 \div 2a^2$ **b** $12a^7 \div 4a^2$ **c** $25a^6 \div 5a$ **d** $48a^8 \div 6a^{-1}$

e $24a^6 \div 8a^{-2}$ **f** $36a \div 6a^5$

7   Simplify these expressions.

a $3a^3b^2 \times 4a^3b$ **b** $7a^3b^5 \times 2ab^3$ **c** $4a^3b^5 \times 5a^4b^{-1}$

d $12a^3b^5 \div 4ab$ **e** $24a^3b^5 \div 6a^2b^{-3}$

8   **a**   Write down two possible multiplication questions with an answer of $18x^3y^4$.

b   Write down two possible division questions with an answer of $18x^3y^4$.

PS 9 a and b are different prime numbers.

What is the smallest value of a^2b^2?

PS 10   Use the general rule for dividing powers of the same number

$$a^x \div a^y = a^{x-y}$$

to prove that any number raised to the power -1 is the reciprocal of that number.

HOMEWORK 5D

Evaluate the following.

1 $36^{\frac{1}{2}}$ **2** $144^{\frac{1}{2}}$ **3** $25^{\frac{1}{2}}$ **4** $196^{\frac{1}{2}}$ **5** $8^{\frac{1}{3}}$ **6** $125^{\frac{1}{3}}$

7 $32^{-\frac{1}{5}}$ **8** $144^{-\frac{1}{2}}$ **9** $27^{-\frac{1}{3}}$ **10** $(\frac{25}{81})^{\frac{1}{2}}$ **11** $(\frac{81}{36})^{\frac{1}{2}}$ **12** $(\frac{36}{64})^{\frac{1}{2}}$

13 $(\frac{8}{27})^{\frac{1}{3}}$ **14** $(\frac{16}{625})^{\frac{1}{4}}$ **15** $(\frac{4}{9})^{-\frac{1}{2}}$ **16** $(\frac{16}{25})^{-\frac{1}{2}}$ **17** $(\frac{8}{27})^{-\frac{1}{3}}$

AU 18   Which of these is the odd one out?

$27^{-\frac{1}{3}}$ $25^{-\frac{1}{2}}$ 3^{-1}

Show how you decided.

AU 19   Imagine that you are the teacher.

Write down how you would teach the class that $16^{-\frac{1}{4}}$ is equal to $\frac{1}{2}$.

PS 20 $x^{-\frac{1}{4}} = y^{-\frac{1}{2}}$

Find values for x and y that make this equation work.

HOMEWORK 5E

1   Evaluate the following.

a $16^{\frac{3}{4}}$ **b** $125^{\frac{4}{3}}$ **c** $81^{\frac{3}{4}}$

2   Rewrite the following in index form.

a $\sqrt[4]{t^3}$ **b** $\sqrt[5]{m^2}$

3   Evaluate the following.

a $27^{\frac{2}{3}}$ **b** $8^{\frac{4}{3}}$ **c** $36^{\frac{3}{2}}$ **d** $81^{1.25}$

4 Using a trial-and-improvement method, or otherwise, solve these equations.
   a $6^x = 60$ **b** $10^x = 2$ (You could try to use the **power** key on your calculator.)

5 **a** Evaluate $8^{\frac{1}{3}}$. **b** Write $16^{-\frac{1}{2}} \times 2^{-3}$ as a power of 2.
   c Given that $32^y = 2$, find the value of y.

PS 6 Which of these is the odd one out?
   $16^{-\frac{3}{4}}$ $64^{-\frac{1}{2}}$ $8^{-\frac{2}{3}}$
   Show how you decided.

AU 7 Imagine that you are the teacher.
   Write down how you would teach the class that $27^{-\frac{2}{3}}$ is equal to $\frac{1}{9}$.

5.2 Standard form

HOMEWORK 5F

1 Evaluate the following.
   a 3.5×100 **b** 2.15×10 **c** 6.74×1000 **d** 4.63×10
   e 30.145×10 **f** 78.56×1000 **g** 6.42×10^2 **h** 0.067×10
   i 0.085×10^3 **j** 0.798×10^5 **k** 0.658×1000 **l** 215.3×10^2
   m 0.889×10^6 **n** 352.147×10^2 **o** 37.2841×10^3 **p** 34.28×10^6

2 Evaluate the following.
   a $4538 \div 100$ **b** $435 \div 10$ **c** $76459 \div 1000$ **d** $643.7 \div 10$
   e $4228.7 \div 100$ **f** $278.4 \div 1000$ **g** $246.5 \div 10^2$ **h** $76.3 \div 10$
   i $76 \div 10^3$ **j** $897 \div 10^5$ **k** $86.5 \div 1000$ **l** $1.5 \div 10^2$
   m $0.8799 \div 10^6$ **n** $23.4 \div 10^2$ **o** $7654 \div 10^3$ **p** $73.2 \div 10^6$

3 Evaluate the following.
   a 400×300 **b** 50×4000 **c** 70×200 **d** 30×700
   e $(30)^2$ **f** $(50)^3$ **g** $(200)^2$ **h** 40×150
   i 60×5000 **j** 30×250 **k** 700×200

4 Evaluate the following.
   a $4000 \div 800$ **b** $9000 \div 30$ **c** $7000 \div 200$ **d** $8000 \div 200$
   e $2100 \div 700$ **f** $9000 \div 60$ **g** $700 \div 50$ **h** $3500 \div 70$
   i $3000 \div 500$ **j** $30\,000 \div 2000$ **k** $5600 \div 1400$ **l** $6000 \div 30$

5 Evaluate the following.
   a 7.3×10^2 **b** 3.29×10^5 **c** 7.94×10^3 **d** 6.8×10^7
   e $3.46 \div 10^2$ **f** $5.07 \div 10^4$ **g** $2.3 \div 10^4$ **h** $0.89 \div 10^3$

AU 6 The diameter of the planet Venus is approximately 7.5×10^3 miles.
   The diameter of Saturn is approximately 7.5×10^4 miles.
   The diameter of Earth is approximately 7.9×10^3 miles.
   Without working out the answers, explain how you can tell which planet is the biggest of the three.

PS 7 A number is between 10 000 and 100 000. It is written in the form 2.5×10^n.
   What is the value of n?

HOMEWORK 5G

1 Write these standard form numbers out in full.

   a 3.5×10^2 **b** 4.15×10 **c** 5.7×10^{-3} **d** 1.46×10

   e 3.89×10^{-2} **f** 4.6×10^3 **g** 2.7×10^2 **h** 8.6×10

   i 4.6×10^3 **j** 3.97×10^5 **k** 3.65×10^{-3} **l** 7.05×10^2

2 Write these numbers in standard form.

   a 780   **b** 0.435   **c** 67 800   **d** 7 400 000 000

   e 30 780 000 000   **f** 0.000 427 8   **g** 6450   **h** 0.047

   i 0.000 12   **j** 96.43   **k** 74.78   **l** 0.004 157 8

3 Write the appropriate numbers given in each statement in standard form.

   a In 1990 there were 24 673 000 vehicles licensed in the UK.

   b In 2001 Keith Gordon was one of 15 282 runners to complete the Boston Marathon.

   c In 1990 the total number of passenger kilometres on the British roads was 613 000 000 000.

   d The Sun is 93 million miles away from Earth. The next nearest star to the Earth is Proxima Centuri which is about 24 million million miles away.

   e A scientist was working with a new particle reported to weigh only 0.000 000 000 000 65 g.

AU 4 How many times smaller is 1.7×10^2 than 1.7×10^5?

AU 5 How many times bigger is 9.6×10^7 than 4.8×10^3?

PS 6 How many times bigger is 1.2×10^5 than 3000?

PS 7 The speed of light is 3.00×10^8 m/s ($2.997\ 924\ 58 \times 10^8$ m/s to be exact).

It takes about 1.3 seconds for light to travel to the moon.

Use this information to work out the distance to the moon.

Give your answer in kilometres.

HOMEWORK 5H

1 Find the results of the following, leaving your answers in standard form.

   a $(4 \times 10^6) \times (7 \times 10^9)$ **b** $(7 \times 10^5) \times (5 \times 10^7)$

   c $(3 \times 10^{-5}) \times (8 \times 10^8)$ **d** $(2.1 \times 10^7) \times (5 \times 10^{-8})$

2 Find the results of the following, leaving your answers in standard form.

   a $(9 \times 10^8) \div (3 \times 10^4)$ **b** $(2.7 \times 10^7) \div (9 \times 10^3)$

   c $(5.5 \times 10^4) \div (1.1 \times 10^{-2})$ **d** $(4.2 \times 10^{-9}) \div (3 \times 10^{-8})$

3 Find the results of the following, leaving your answers in standard form.

   a $\dfrac{8 \times 10^9}{4 \times 10^7}$ **b** $\dfrac{12 \times 10^6}{3 \times 10^4}$ **c** $\dfrac{2.8 \times 10^7}{7 \times 10^{-4}}$

4 $p = 8 \times 10^5$ and $q = 2 \times 10^7$

Find the value of the following, leaving your answer in standard form.

   a $p \times q$ **b** $p \div q$ **c** $p + q$ **d** $q - p$ **e** $\dfrac{q}{p}$

5 $p = 2 \times 10^{-2}$ and $q = 4 \times 10^{-3}$

    a $p \times q$ **b** $p \div q$ **c** $p + q$ **d** $q - p$ **e** $\dfrac{q}{p}$

FM 6 In 2010 the population of Africa was approximately 1×10^9.
In 2050 the population of Africa is expected to be 1.8×10^9.
By how much is the population of Africa expected to rise?
Give your answer in millions.

AU 7 A number, when written in standard form, is greater than 1 million and less than 5 million.
Write down a possible value of the number in standard form.

PS 8 Here are four numbers written in standard form.

    3.5×10^5 1.2×10^3 7.3×10^2 4.8×10^4

    a Work out the largest answer when two of these numbers are multiplied together.
    b Work out the smallest answer when two of these numbers are added together.
Give your answers in standard form.

5.3 Rational numbers and reciprocals

HOMEWORK 5I

1 Work out each of these fractions as a decimal. Give them as terminating decimals or recurring decimals as appropriate.

    a $\dfrac{3}{4}$ **b** $\dfrac{1}{15}$ **c** $\dfrac{1}{25}$ **d** $\dfrac{1}{11}$ **e** $\dfrac{1}{20}$

PS 2 There are several patterns to be found in recurring decimals. For example,

    $\dfrac{1}{13} = 0.076923076923076923076923\ldots$; $\dfrac{2}{13} = 0.153846153846153846153846\ldots$
    $\dfrac{3}{13} = 0.230769230769230769230769\ldots$ and so on

    a Write down the decimals for $\dfrac{4}{13}, \dfrac{5}{13}, \dfrac{6}{13}, \dfrac{7}{13}, \dfrac{8}{13}, \dfrac{9}{13}, \dfrac{10}{13}, \dfrac{11}{13}, \dfrac{12}{13}$ to 24 decimal places.
    b What do you notice?

3 Write each of these fractions as a decimal. Use this to write the list in order of size, smallest first.

    $\dfrac{2}{9}$ $\dfrac{1}{5}$ $\dfrac{23}{100}$ $\dfrac{2}{7}$ $\dfrac{3}{11}$

4 Convert each of these terminating decimals to a fraction.

    a 0.57   **b** 0.275   **c** 0.85   **d** 0.06   **e** 3.65

AU 5 Explain why the reciprocal of 1 is 1.

PS 6 **a** Work out the reciprocal of the reciprocal of 4.
    b Work out the reciprocal of the reciprocal of 5.
    c What do you notice?

7 **a** Give an example to show that the reciprocal of a number greater than 1 is less than 1.
    b Give an example to show that the reciprocal of a number less than 0 is also less than 0.

8 $x = 0.0242\,424\ldots$

    a What is $100x$?
    b By subtracting the original value from your answer to part **a**, work out the value of $99x$.
    c Multiply both sides by 10 to get $990x$ and eliminate the decimal on the right-hand side.
    d Divide both sides by 990.
    e What is x as a fraction expressed in its lowest terms?

Convert each of these recurring decimals to a fraction.

a $0.\dot{7}$ **b** $0.5\dot{7}$ **c** $0.5\dot{4}$ **d** $0.\dot{2}7\dot{5}$

e $2.\dot{5}$ **f** $2.\dot{3}\dot{6}$ **g** $0.06\dot{3}$ **h** $2.07\dot{5}$

10   **a**   Write 1.7 as a rational number in the form $\frac{a}{b}$, where a and b are whole numbers.

     b   Given that $n = 1.\dot{7}$

         i   Write down the value of $10n$.

         ii   Hence write down the value of $9n$.

         iii   Express n as a rational number, in the form $\frac{a}{b}$, where a and b are whole numbers.

5.4 Surds

HOMEWORK 5J

1   Work out each of the following in simplified form.

     a $\sqrt{3} \times \sqrt{4}$ **b** $\sqrt{5} \times \sqrt{7}$ **c** $\sqrt{5} \times \sqrt{5}$ **d** $\sqrt{2} \times \sqrt{32}$

PS 2   Work out each of the following in surd form.

     a $\sqrt{15} \div \sqrt{5}$ **b** $\sqrt{18} \div \sqrt{2}$ **c** $\sqrt{32} \div \sqrt{2}$ **d** $\sqrt{12} \div \sqrt{8}$

3   Work out each of the following in surd form.

     a $\sqrt{3} \times \sqrt{3} \times \sqrt{2}$ **b** $\sqrt{5} \times \sqrt{5} \times \sqrt{15}$ **c** $\sqrt{2} \times \sqrt{8} \times \sqrt{8}$ **d** $\sqrt{2} \times \sqrt{8} \times \sqrt{5}$

4   Work out each of the following in surd form.

     a $\sqrt{3} \times \sqrt{8} \div \sqrt{2}$ **b** $\sqrt{15} \times \sqrt{3} \div \sqrt{5}$ **c** $\sqrt{8} \times \sqrt{8} \div \sqrt{2}$ **d** $\sqrt{3} \times \sqrt{27} \div \sqrt{3}$

5   Simplify each of the following surds into the form $a\sqrt{b}$.

     a $\sqrt{90}$ **b** $\sqrt{32}$ **c** $\sqrt{63}$ **d** $\sqrt{300}$

     e $\sqrt{150}$ **f** $\sqrt{270}$ **g** $\sqrt{96}$ **h** $\sqrt{125}$

6   Simplify each of these.

     a $2\sqrt{32} \times 5\sqrt{2}$ **b** $4\sqrt{8} \times 2\sqrt{2}$ **c** $4\sqrt{12} \times 5\sqrt{3}$ **d** $3\sqrt{6} \times 2\sqrt{6}$

     e $2\sqrt{5} \times 5\sqrt{3}$ **f** $2\sqrt{3} \times 3\sqrt{3}$ **g** $2\sqrt{2} \times 3\sqrt{8}$ **h** $2\sqrt{3} \times 2\sqrt{27}$

     i $8\sqrt{24} \div 2\sqrt{3}$ **j** $3\sqrt{27} \div \sqrt{3}$ **k** $5\sqrt{18} \div \sqrt{2}$ **l** $2\sqrt{32} \div 4\sqrt{8}$

     m $5\sqrt{2} \times \sqrt{8} \div 2\sqrt{2}$ **n** $3\sqrt{15} \times \sqrt{3} \div \sqrt{5}$ **o** $2\sqrt{24} \times 5\sqrt{3} \div 2\sqrt{8}$

PS 7   Find the value of a that makes each of these surds true.

     a $\sqrt{5} \times \sqrt{a} = 20$ **b** $\sqrt{3} \times \sqrt{a} = 12$ **c** $\sqrt{5} \times 4\sqrt{a} = 20$

8   Simplify the following.

     a $\left(\frac{\sqrt{2}}{3}\right)^2$ **b** $\left(\frac{4}{\sqrt{3}}\right)^2$

9   Simplify the following.

     a $\sqrt{32} + \sqrt{8}$ **b** $\sqrt{32} \times \sqrt{8}$ **c** $\sqrt{27} \times \sqrt{18} \div \sqrt{3}$

AU 10   Decide whether this statement is true or false.

     $\sqrt{(a^2 + b^2)} = a + b$

     Show your working.

PS 11   Write down a division of two different surds which has an integer answer.

HOMEWORK 5K

1 Rationalise the denominators of these expressions.

a $\dfrac{1}{\sqrt{7}}$ **b** $\dfrac{1}{\sqrt{8}}$ **c** $\dfrac{2}{\sqrt{5}}$ **d** $\dfrac{1}{2\sqrt{2}}$

e $\dfrac{5\sqrt{3}}{\sqrt{27}}$ **f** $\dfrac{\sqrt{8}}{\sqrt{3}}$ **g** $\dfrac{1+\sqrt{3}}{\sqrt{3}}$ **h** $\dfrac{3-\sqrt{2}}{\sqrt{8}}$

2 Show that:

a $(3+\sqrt{5})(2+\sqrt{5}) = 11 + 5\sqrt{5}$ **b** $(3-\sqrt{2})(3+\sqrt{2}) = 7$

3 Expand and simplify where possible.

a $\sqrt{5}(3-\sqrt{2})$ **b** $\sqrt{8}(3-4\sqrt{2})$ **c** $3\sqrt{8}(2\sqrt{2}+4)$
d $(2+\sqrt{3})(1-\sqrt{3})$ **e** $(3+\sqrt{5})(2-\sqrt{5})$ **f** $(3-\sqrt{2})(4+2\sqrt{2})$

4 Work out the missing lengths in these triangles, simplifying the answer where possible.

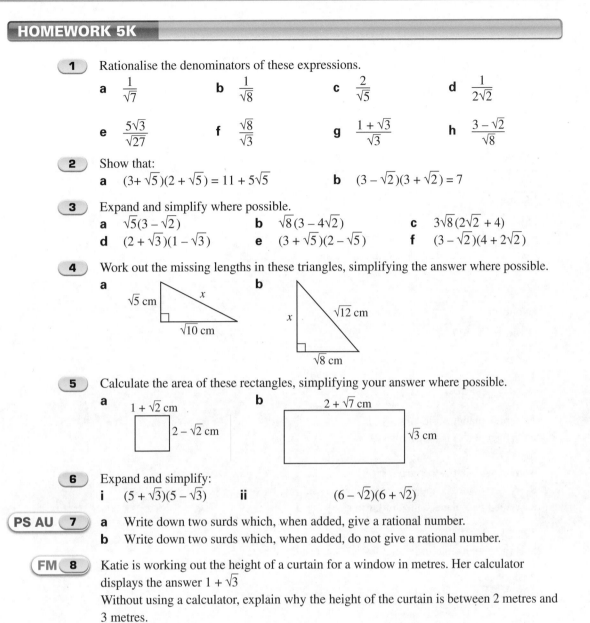

a
$\sqrt{5}$ cm $\quad x$
$\sqrt{10}$ cm

b
$x \quad \sqrt{12}$ cm
$\sqrt{8}$ cm

5 Calculate the area of these rectangles, simplifying your answer where possible.

a
$1 + \sqrt{2}$ cm
$2 - \sqrt{2}$ cm

b
$2 + \sqrt{7}$ cm
$\sqrt{3}$ cm

6 Expand and simplify:

i $(5+\sqrt{3})(5-\sqrt{3})$ **ii** $(6-\sqrt{2})(6+\sqrt{2})$

PS AU 7 **a** Write down two surds which, when added, give a rational number.
b Write down two surds which, when added, do not give a rational number.

FM 8 Katie is working out the height of a curtain for a window in metres. Her calculator displays the answer $1 + \sqrt{3}$

Without using a calculator, explain why the height of the curtain is between 2 metres and 3 metres.

A*

Functional Maths Activity

The planets

The table shows information about the planets in our solar system in order of distance from the sun.

Planet	Distance from the sun (million km)	Mass (kg)	Diameter (km)
Mercury	58	3.3×10^{23}	4878
Venus	108	4.87×10^{24}	12 104
Earth	150	5.98×10^{24}	12 756
Mars	228	6.42×10^{23}	6787
Jupiter	778	1.90×10^{27}	142 796
Saturn	1427	5.69×10^{26}	120 660
Uranus	2871	8.68×10^{25}	51 118
Neptune	4497	1.02×10^{26}	48 600
Pluto	5913	1.29×10^{22}	2274

Task 1

Answer the following questions.

1 Which planet is the largest?
2 Which planet is the smallest?
3 Which planet is the lightest?
4 Which planet is the heaviest?
5 Which planet is approximately twice as far away from the sun as Saturn?
6 Which two planets are similar in size?

Task 2

Sort the planets into order, lightest to heaviest, by mass.

Task 3

Sort the planets into order, shortest to longest, by diameter.

Task 4

Imagine you are a scientist working for NASA. Can you identify a relationship between each planet's mass, distance from the sun and diameter? Are there any trends? Write your findings in a brief report.

Algebra: Quadratic equations

6.1 Expanding brackets

HOMEWORK 6A

Expand the following expressions.

1 $(x + 2)(x + 5)$ **2** $(t + 3)(t + 2)$ **3** $(w + 4)(w + 1)$

4 $(m + 6)(m + 2)$ **5** $(k + 2)(k + 4)$ **6** $(a + 3)(a + 1)$

7 $(x + 3)(x - 1)$ **8** $(t + 6)(t - 4)$ **9** $(w + 2)(w - 3)$

10 $(f + 1)(f - 4)$ **11** $(g + 2)(g - 5)$ **12** $(y + 5)(y - 2)$

13 $(x - 4)(x + 3)$ **14** $(p - 3)(p + 2)$ **15** $(k - 5)(k + 1)$

16 $(y - 3)(y + 6)$ **17** $(a - 2)(a + 4)$ **18** $(t - 4)(t + 5)$

19 $(x - 3)(x - 2)$ **20** $(r - 4)(r - 1)$ **21** $(m - 1)(m - 7)$

22 $(g - 5)(g - 3)$ **23** $(h - 6)(h - 2)$ **24** $(n - 2)(n - 8)$

25 $(4 + x)(3 + x)$ **26** $(5 + t)(4 - t)$ **27** $(2 - b)(6 + b)$

28 $(7 - y)(5 - y)$ **29** $(3 + p)(p - 2)$ **30** $(3 - k)(k - 5)$

PS 31 This rectangle is made up of four parts with areas of x^2, $5x$, $4x$ and 20 square units.

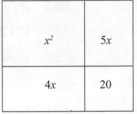

Work out expressions for the sides of the rectangle in terms of x.

PS 32 This square has an area of x^2 square units.

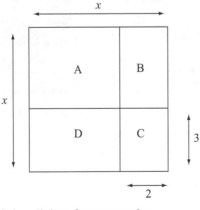

It is split into four rectangles.

a Fill in the table below to show the dimensions and area of each rectangle.

Rectangle	Length	Height	Area
A	$x - 2$	$x - 3$	$(x - 2)(x - 3)$
B			
C			
D			

b Add together the areas of rectangles B, C and D.
Expand any brackets and collect the terms.

c Use the results to explain why $(x - 2)(x - 3) = x^2 - 5x + 6$.

AU 33 **a** Expand $(x - 2)(x + 2)$.

b Use the result from **a** to write down the answers to the following (do not use a calculator or do a long multiplication):
  i 98×102 **ii** 198×202

HOMEWORK 6B

Expand the following expressions.

1 $(3x + 4)(4x + 2)$ **2** $(2y + 1)(3y + 2)$ **3** $(4t + 2)(3t + 6)$

4 $(3t + 2)(2t - 1)$ **5** $(6m + 1)(3m - 2)$ **6** $(5k + 3)(4k - 3)$

7 $(4p - 5)(3p + 4)$ **8** $(6w + 1)(3w + 4)$ **9** $(3a - 4)(5a + 1)$

10 $(5r - 2)(3r - 1)$ **11** $(4g - 1)(3g - 2)$ **12** $(3d - 2)(4d + 1)$

13 $(3 + 4p)(5 + 4p)$ **14** $(3 + 2t)(5 + 3t)$ **15** $(2 + 5p)(3p + 1)$

16 $(7 + 4t)(3 - 2t)$ **17** $(5 + 2n)(4 - n)$ **18** $(3 + 4f)(5f - 1)$

19 $(2 - 3q)(5 + 4q)$ **20** $(3 - p)(2 + 3p)$ **21** $(5 - 3t)(4t + 1)$

22 $(5 - 4r)(3 - 4r)$ **23** $(4 - x)(1 - 5x)$ **24** $(2 - 7m)(2m - 3)$

25 $(x + y)(3x + 5y)$ **26** $(4y + t)(3y - 4t)$ **27** $(5x - 3y)(5x + y)$

28 $(x - 2y)(x - 3y)$ **29** $(4m - 3p)(m + 5p)$ **30** $(t - 4k)(3t - k)$

PS 31 **a** Expand $(x + 1)(x + 1)$

b Expand $(x - 1)(x - 1)$

c Expand $(x + 1)(x - 1)$

d Use the results in parts **a**, **b** and **c** to show that $(p - q)^2 \equiv p^2 - 2pq + q^2$ is an identity.

> **HINTS AND TIPS**
>
> Take $p = x + 1$ and $q = x - 1$.

AU 32 **a** Without expanding the brackets, match each expression on the left with an expression on the right. One has been done for you.

$(4x - 3)(3x + 2)$ $12x^2 - x - 1$
$(2x - 1)(6x - 1)$ $12x^2 - x - 6$
$(6x - 1)(2x + 3)$ $12x^2 + 37x + 3$
$(4x + 1)(3x - 1)$ $12x^2 - 8x + 1$
$(x + 3)(12x + 1)$ $12x^2 + 16x - 3$

b Taking any expression on the left, explain how you can match it with an expression on the right without expanding the brackets.

HOMEWORK 6C

Try to spot the pattern in each of the following expressions so that you can immediately write down the expansion.

1 $(x + 1)(x - 1)$ **2** $(t + 2)(t - 2)$ **3** $(y + 3)(y - 3)$

4 $(2m + 3)(2m - 3)$ **5** $(4k - 3)(4k + 3)$ **6** $(5h - 1)(5h + 1)$

7 $(3 + 2x)(3 - 2x)$ **8** $(7 + 2t)(7 - 2t)$ **9** $(4 - 5y)(4 + 5y)$

10 $(a + b)(a - b)$ **11** $(3t + k)(3t - k)$ **12** $(m - 3p)(m + 3p)$

13 $(8k + g)(8k - g)$ **14** $(ac + bd)(ac - bd)$ **15** $(x^2 + y^2)(x^2 - y^2)$

PS 16 Imagine a square of side $2a$ units with a square of side b units cut from one corner.

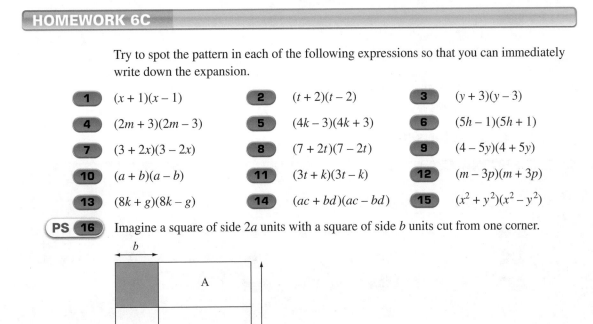

a What does the remaining area become after the small square is cut away?

b The remaining area is cut into three rectangles, A, B and C, and is rearranged as shown here.
Write down the dimensions and area of the rectangle formed by A, B and C.

c Explain why $4a^2 - b^2 = (2a + b)(2a - b)$.

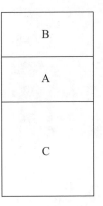

HOMEWORK 6D

Expand the following squares.

1 $(x + 4)^2$ **2** $(m + 3)^2$ **3** $(5 + t)^2$ **4** $(2 + p)^2$

5 $(m - 2)^2$ **6** $(t - 4)^2$ **7** $(3 - m)^2$ **8** $(6 - k)^2$

9 $(2x + 1)^2$ **10** $(3t + 2)^2$ **11** $(1 + 4y)^2$ **12** $(2 + m)^2$

13 $(3t - 2)^2$ **14** $(2x - 1)^2$ **15** $(1 - 4t)^2$ **16** $(5 - 4r)^2$

17 $(a + b)^2$ **18** $(x - y)^2$ **19** $(3t + y)^2$ **20** $(m - 2n)^2$

21 $(x + 3)^2 - 4$ **22** $(x - 4)^2 - 25$ **23** $(x + 5)^2 - 36$ **24** $(x - 1)^2 - 1$

PS 25 A teacher asks her class to expand $(4x - 1)^2$.

Bernice's answer is $16x^2 - 1$

Piotr's answer is $16x^2 + 8x - 1$

a Explain the mistakes that Bernice has made.

b Explain the mistakes that Piotr has made.

c Work out the correct answer.

6.2 Quadratic factorisation

HOMEWORK 6E

Factorise the following.

1 $x^2 + 7x + 6$ **2** $t^2 + 4t + 4$ **3** $m^2 + 11m + 10$ **4** $k^2 + 11k + 24$

5 $p^2 + 10p + 24$ **6** $r^2 + 11r + 18$ **7** $w^2 + 9w + 18$ **8** $x^2 + 8x + 12$

9 $a^2 + 13a + 12$ **10** $k^2 - 10k + 21$ **11** $f^2 - 22f + 21$ **12** $b^2 + 35b + 96$

13 $t^2 + 5t + 6$ **14** $m^2 - 5m + 4$ **15** $p^2 - 7p + 10$ **16** $x^2 - 13x + 36$

17 $c^2 - 12c + 32$ **18** $t^2 - 15t + 36$ **19** $y^2 - 14y + 48$ **20** $j^2 - 19j + 48$

21 $p^2 + 8p + 15$ **22** $y^2 + y - 6$ **23** $t^2 + 7t - 8$ **24** $x^2 + 9x - 10$

25 $m^2 - m - 12$ **26** $r^2 + 6r - 7$ **27** $n^2 - 7n - 18$ **28** $m^2 - 20m - 44$

29 $w^2 - 5w - 24$ **30** $t^2 + t - 90$ **31** $x^2 - x - 72$ **32** $t^2 - 18t - 63$

33 $d^2 - 2d + 1$ **34** $y^2 + 29y + 100$ **35** $t^2 - 10t + 16$ **36** $m^2 - 30m + 81$

37 $x^2 - 30x + 144$ **38** $d^2 - 4d - 12$ **39** $t^2 + t - 20$ **40** $q^2 + q - 56$

41 $p^2 - p - 2$ **42** $v^2 - 2v - 35$ **43** $t^2 - 4t + 3$ **44** $m^2 + 3m - 4$

PS 45 The rectangle below is made up of four parts.

Two of the parts have areas of x^2 and nine square units.

The sides of the rectangle are of the form $x + a$ and $x + b$.

There are two possible answers for a and b.

Work out both answers and copy and complete the areas in the other parts of the rectangle.

AU 46 **a** Expand $(x - a)(x - b)$

b If $x^2 - 9x + 18 = (x - p)(x - q)$, use your answer to part **a** to write down the values of:

    i $p + q$ **ii** pq

c Explain how you can tell that $x^2 - 18x + 9$ will not factorise.

HOMEWORK 6F

Each of these is the difference of two squares. Factorise them.

1 $x^2 - 81$ **2** $t^2 - 36$ **3** $4 - x^2$ **4** $81 - t^2$

5 $k^2 - 400$ **6** $64 - y^2$ **7** $x^2 - y^2$

8 $a^2 - 9b^2$ **9** $9x^2 - 25y^2$ **10** $9x^2 - 16$ **11** $100t^2 - 4w^2$

12 $36a^2 - 49b^2$

13 Simplify: $\dfrac{2a - 3}{4a^2 - 9}$

PS 14
a A square has a side of $2x$ units.
What is the area of the square?

b A rectangle (A), three units wide, is cut from the square and placed at the side of the remaining rectangle (B).

A square (C) is then cut from the bottom of rectangle A to leave a final rectangle (D), as shown in the diagram.

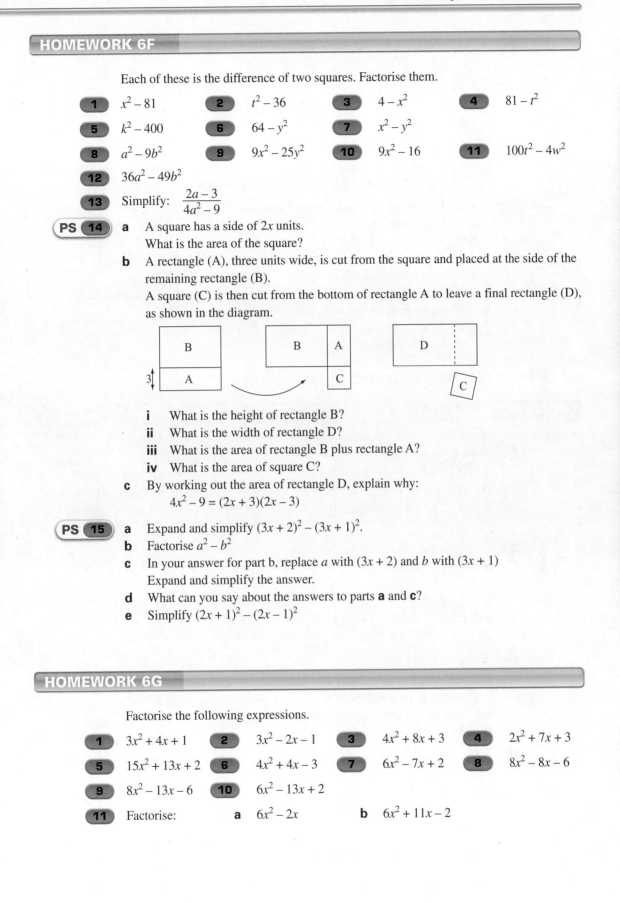

i What is the height of rectangle B?
ii What is the width of rectangle D?
iii What is the area of rectangle B plus rectangle A?
iv What is the area of square C?

c By working out the area of rectangle D, explain why:
$4x^2 - 9 = (2x + 3)(2x - 3)$

PS 15
a Expand and simplify $(3x + 2)^2 - (3x + 1)^2$.
b Factorise $a^2 - b^2$
c In your answer for part b, replace a with $(3x + 2)$ and b with $(3x + 1)$
Expand and simplify the answer.
d What can you say about the answers to parts **a** and **c**?
e Simplify $(2x + 1)^2 - (2x - 1)^2$

HOMEWORK 6G

Factorise the following expressions.

1 $3x^2 + 4x + 1$ **2** $3x^2 - 2x - 1$ **3** $4x^2 + 8x + 3$ **4** $2x^2 + 7x + 3$

5 $15x^2 + 13x + 2$ **6** $4x^2 + 4x - 3$ **7** $6x^2 - 7x + 2$ **8** $8x^2 - 8x - 6$

9 $8x^2 - 13x - 6$ **10** $6x^2 - 13x + 2$

11 Factorise: **a** $6x^2 - 2x$ **b** $6x^2 + 11x - 2$

PS 12 This rectangle is made up of four parts, with areas of $6x^2$, $3x$, $8x$ and four square units.

$6x^2$	$3x$
$8x$	4

Work out expressions for the sides of the rectangle in terms of x.

AU 13 Three pupils are asked to factorise the expression $4x^2 + 4x - 8$

These are their answers:

Adam   Bertie   Cara
$(2x + 2)(2x - 2)$ $(4x + 8)(x - 1)$ $(x + 2)(4x - 4)$

All the answers are correctly factorised.

a Explain why one quadratic expression can have three different factorisations.

b Which of the following is the most complete factorisation?

$2(x + 2)(2x - 2)$ $4(x + 2)(x - 1)$ $2(2x + 4)(x - 1)$

Explain your choice.

6.3 Solving quadratic equations by factorisation

HOMEWORK 6H

1 Solve these equations.

a $(x + 3)(x + 2) = 0$ **b** $(t + 4)(t + 1) = 0$ **c** $(a + 5)(a + 3) = 0$

d $(x + 4)(x - 1) = 0$ **e** $(x + 2)(x - 5) = 0$ **f** $(t + 3)(t - 4) = 0$

g $(x - 2)(x + 1) = 0$ **h** $(x - 1)(x + 4) = 0$ **i** $(a - 6)(a + 5) = 0$

j $(x - 2)(x - 5) = 0$ **k** $(x - 2)(x - 1) = 0$ **l** $(a - 2)(a - 6) = 0$

2 First factorise, then solve these equations.

a $x^2 + 6x + 5 = 0$ **b** $x^2 + 9x + 18 = 0$ **c** $x^2 - 7x + 8 = 0$

d $x^2 - 4x - 21 = 0$ **e** $x^2 + 3x - 10 = 0$ **f** $x^2 + 2x - 15 = 0$

g $t^2 - 4t - 12 = 0$ **h** $t^2 - 3t - 18 = 0$ **i** $x^2 + x - 2 = 0$

j $x^2 - 4x + 4 = 0$ **k** $m^2 - 10m + 25 = 0$ **l** $t^2 - 10t + 16 = 0$

m $t^2 + 7t + 12 = 0$ **n** $k^2 - 3k - 18 = 0$ **o** $a^2 - 20a + 64 = 0$

PS 3 A woman is x years old.

Her brother is four years older than her.

The product of their ages is 1020.

a Set up a quadratic equation to represent this situation.

b How old is the woman?

FM 4 A rectangular field is 140 m longer than it is wide.

A combine harvester cuts corn at the rate of 200 m^2 per minute.

It takes four hours to cut the field.

What are the dimensions of the field?

HOMEWORK 6I

1 Give your answers either in rational form or as mixed numbers.

   a Solve the following equations.

     i $2x^2 + 5x + 2 = 0$ **ii** $7x^2 + 8x + 1 = 0$ **iii** $4x^2 + 3x - 7 = 0$

     iv $6x^2 + 13x + 5 = 0$ **v** $6x^2 + 7x + 2 = 0$

   b Solve the following equations.

     i $x^2 - x = 6$ **ii** $2x(4x + 7) = -3$ **iii** $(x + 3)(x - 4) = 18$

     iv $11x = 21 - 2x^2$ **v** $(2x + 3)(2x - 3) = 9x$

   c **i**   Simplify $\dfrac{x^2 - 9}{3x - 9}$ **ii**   Solve the equation $12x^2 - 25x + 12 = 0$

AU 2 Here are three equations.

   A: $(x - 2)^2 = 0$ B: $3x - 2 = 4$ C: $x^2 - 4x + 4 = 0$

   a Give a mathematical fact that equations A and C have in common.

   b Give a mathematical fact that equations A, B and C have in common.

PS 3 Pythagoras' theorem states that the sum of the squares of the two short sides of a right-angled triangle equals the square of the long side (hypotenuse).

A right-angled triangle has sides $4x$, $2 - x$ and $x + 1$ cm.

   a Show that $2x^2 + x - 1 = 0$

   b Find the area of the triangle.

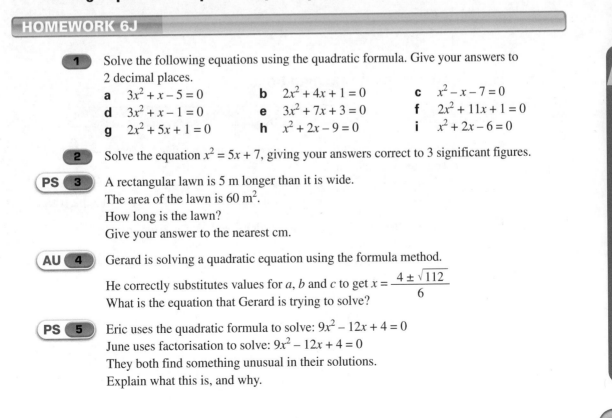

6.4 Solving a quadratic equation by the quadratic formula

HOMEWORK 6J

1 Solve the following equations using the quadratic formula. Give your answers to 2 decimal places.

   a $3x^2 + x - 5 = 0$ **b** $2x^2 + 4x + 1 = 0$ **c** $x^2 - x - 7 = 0$

   d $3x^2 + x - 1 = 0$ **e** $3x^2 + 7x + 3 = 0$ **f** $2x^2 + 11x + 1 = 0$

   g $2x^2 + 5x + 1 = 0$ **h** $x^2 + 2x - 9 = 0$ **i** $x^2 + 2x - 6 = 0$

2 Solve the equation $x^2 = 5x + 7$, giving your answers correct to 3 significant figures.

PS 3 A rectangular lawn is 5 m longer than it is wide.

The area of the lawn is 60 m².

How long is the lawn?

Give your answer to the nearest cm.

AU 4 Gerard is solving a quadratic equation using the formula method.

He correctly substitutes values for a, b and c to get $x = \dfrac{4 \pm \sqrt{112}}{6}$

What is the equation that Gerard is trying to solve?

PS 5 Eric uses the quadratic formula to solve: $9x^2 - 12x + 4 = 0$

June uses factorisation to solve: $9x^2 - 12x + 4 = 0$

They both find something unusual in their solutions.

Explain what this is, and why.

6.5 Solving a quadratic equation by completing the square

HOMEWORK 6K

1 Write an equivalent expression in the form $(x \pm a)^2 - b$
a $x^2 + 10x$ **b** $x^2 + 18x$ **c** $x^2 - 8x$ **d** $x^2 + 20x$ **e** $x^2 + 7x$

2 Write an equivalent expression in the form $(x \pm a)^2 - b$
a $x^2 + 10x - 1$ **b** $x^2 + 18x - 5$ **c** $x^2 - 8x + 3$ **d** $x^2 - 5x - 1$

3 Solve the following equations by completing the square. Leave your answers in surd form where appropriate. The answers to Question **2** will help.
a $x^2 + 10x - 1 = 0$ **b** $x^2 + 18x - 5 = 0$ **c** $x^2 - 8x + 3 = 0$
d $x^2 + 20x + 7 = 0$ **e** $x^2 - 5x - 1 = 0$

4 Solve $x^2 + 8x - 3 = 0$ by completing the square. Give your answers to 2 dp.

5 **a** Write the equation $x^2 + 4x - 6$ in the form $(x + a)^2 - b$
b Hence or otherwise, solve the equation $x^2 + 4x - 6 = 0$, leaving your answer in surd form.

AU 6 Dave rewrites the expression $x^2 + px + q$ using completing the square.
He correctly does this and gets $(x + 3)^2 - 17$
What are the values of p and q?

PS 7 Rearrange the following to give a logical solution to the equation:
$$x^2 + 12x - 11 = 0$$
A: $x = -6 \pm \sqrt{47}$
B: $(x + 6)^2 - 47 = 0$
C: $= (x + 6)^2 - 47$
D: $= (x + 6)^2 - 36 - 11$
E: $(x + 6)^2 = 47$
F: $x + 6 = \pm \sqrt{47}$

6.6 Problems involving quadratic equations

HOMEWORK 6L

1 Work out the discriminant $b^2 - 4ac$ of the following equations. In each case say how many solutions the equation has.
a $3x^2 + 6x + 3 = 0$ **b** $2x^2 + 3x - 5 = 0$ **c** $2x^2 + 3x + 5 = 0$
d $8x^2 + 3x - 2 = 0$ **e** $5x^2 + 4x + 1 = 0$ **f** $4x^2 + 4x + 1 = 0$

PS 2 Bill works out the discriminant of the quadratic equation $x^2 + bx - c = 0$
as: $b^2 - 4ac = 33$
There are six possible equations that could lead to this discriminant, where a, b and c are integers. What are they?

HOMEWORK 6M

PS 1 The sides of a right-angled triangle are $3x$, $(x + 1)$ and $(4x - 3)$. Find the actual dimensions of the triangle.

PS 2 The length of a rectangle is 3 m more than its width. Its area is 130 m². Find the actual dimensions of the rectangle.

3 Solve the equation: $x + \dfrac{2}{x} = 5$

Give your answers correct to 2 decimal places.

4 Solve the equation: $3x + \dfrac{2}{x} = 7$

PS 5 The area of a triangle is 24 cm². The base is 8 cm longer than the height. Use this information to set up a quadratic equation. Solve the equation to find the length of the base.

FM 6 On a journey of 210 km, the driver of a train calculates that if he were to increase his average speed by 10 km/h, he would take 30 minutes less. Find his average speed.

FM 7 After a 25p per kilogram increase in the price of bananas, I can buy 2 kilograms less for £6 than I could last week. How much do bananas cost this week?

8 Gareth took part in a 26-mile race.
- **a** He ran the first 15 miles at an average speed of x mph. He ran the last 11 miles at an average speed of $(x - 2)$ mph. Write down an expression, in terms of x, for the time he took to complete the 26-mile race.
- **b** Gareth took four hours to complete the race. Using your answer to part **a**, form an equation in terms of x.
- **c** **i** Simplify your equation and show that it can be written as: $2x^2 - 17x + 15 = 0$
  - **ii** Solve the equation and obtain Gareth's average speed over the first 15 miles.

FM 9 Ana, an interior decorator, is told that a rectangular room is 4 m longer than it is wide. She is also told that it cost £195 to carpet the room.
cost of the carpet was £12 per square metre.
Help her to work out the width of the room.

Problem-solving Activity

Quadratic equations

1 The sides of a rectangle are x cm and $2x$ cm.

Alex notices that the area of the rectangle in cm² is 80 more than the perimeter in cm.

What size is the rectangle?

2 Suppose the sides of the rectangle in question are x cm and $3x$ cm.

What size is the rectangle in this case?

7.1 Similar triangles

HOMEWORK 7A

1   These diagrams are drawn to scale. What is the scale factor of the enlargement in each case?

a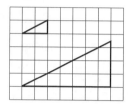

b

2   **a**   Explain why these two shapes are similar.
   b   Give the ratio of the sides.
   c   Which angle corresponds to angle C?
   d   Which side corresponds to side QP?

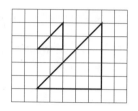

3   In the diagrams below, find the lengths of the sides marked x.
   Each pair of shapes is similar but not drawn to scale.

a

b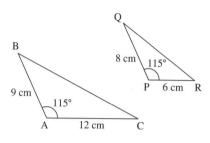

FM 4   Zahid wants to make a frame for his picture. The picture measures 12 cm by 8 cm.

The wood that Zahid is using to make the frame is 10 cm wide.
What length of wood does Zahid need to make the frame, if the picture and the frame are similar.

FM Functional Maths   **AU** (AO2) Assessing Understanding   **PS** (AO3) Problem Solving

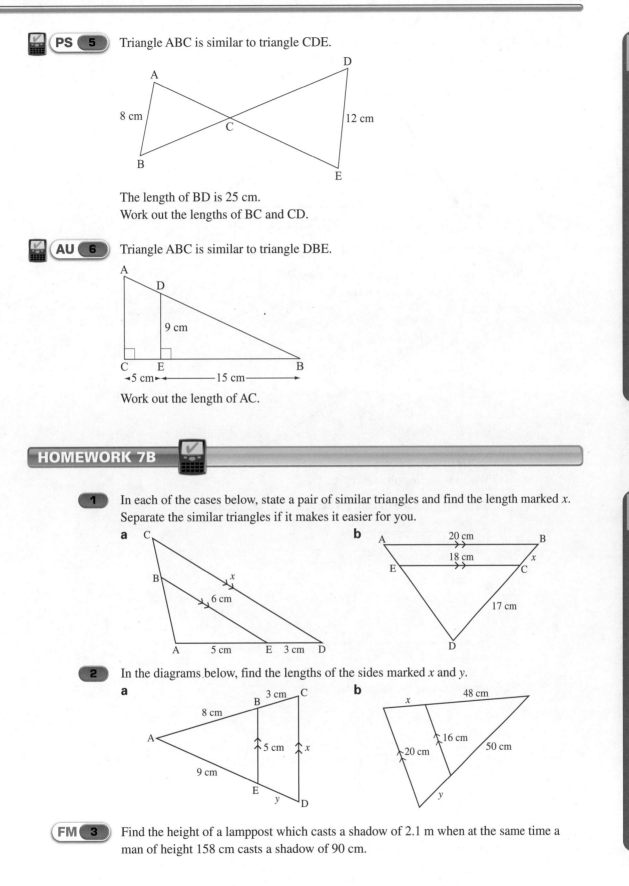

PS 5 Triangle ABC is similar to triangle CDE.

The length of BD is 25 cm.
Work out the lengths of BC and CD.

AU 6 Triangle ABC is similar to triangle DBE.

Work out the length of AC.

HOMEWORK 7B

1 In each of the cases below, state a pair of similar triangles and find the length marked x.
Separate the similar triangles if it makes it easier for you.

a

b

2 In the diagrams below, find the lengths of the sides marked x and y.

a

b

FM 3 Find the height of a lamppost which casts a shadow of 2.1 m when at the same time a
man of height 158 cm casts a shadow of 90 cm.

B

FM 4 Jamie has designed this metal framework for a garden slide.
Triangles ABC and ADE are similar.

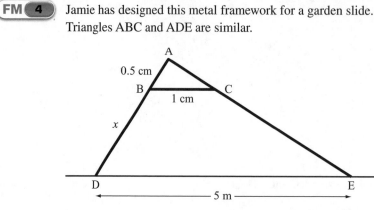

Work out the length of metal he needs for BD, marked x on the diagram.

PS 5 A square ABCD fits inside a triangle DEF.
BE = 10 cm and AE = 6 cm

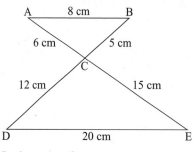

Work out the length of BF.

AU 6 Suzie says that the triangles ABC and CDE are similar.

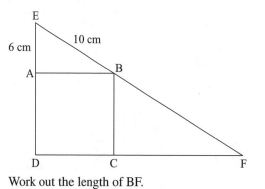

Is she correct?
Show working to explain your answer.

HOMEWORK 7C

1. Find the lengths marked x in the diagrams below.

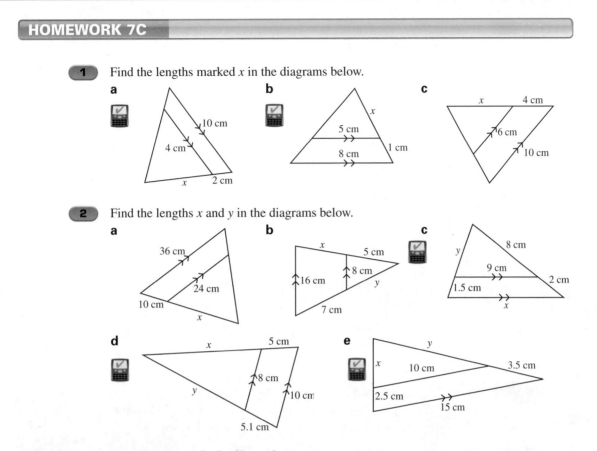

a

b

c

2. Find the lengths x and y in the diagrams below.

a

b

c

d

e

7.2 Areas and volumes of similar shapes

HOMEWORK 7D

1. The length ratio between two similar solids is $3 : 7$.
   a   What is the area ratio between the solids?
   b   What is the volume ratio between the solids?

2. Copy and complete this table.

Linear scale factor	Linear ratio	Linear fraction	Area scale factor	Volume scale factor
4	$1 : 4$	$\frac{4}{1}$		
$\frac{1}{2}$				
	$10 : 1$			
			36	
				125

FM 3   Don wants to do some planting in his garden. He marks out a shape with an area of 20 cm^2. However, he then decides that he wants to use more garden space. What is the area of a similar shape of which the lengths are four times the corresponding lengths of the first shape?

4. A bricklayer uses bricks, each of which has a volume of 400 cm^3. He is trying to decide if he should use a different type of brick. What would be the volume of a similar brick, with lengths of:
   a   three times the corresponding lengths of the first brick
   b   five times the corresponding lengths of the first brick?

A

 A can of paint, 12 cm high, holds five litres of paint. Help Anita to work out how much paint would go into a similar can that is 30 cm high.

 A sculptor has made a model statue that is 15 cm high and has a volume of 450 cm³.
The real statue will be 4.5 m high.
In order to buy enough materials, she needs to know the volume of the real statue.
Work this out for her, giving your answer in m³.

 Tim has a large tin full of paint that he wants to empty into a number of smaller tins.
The diagram shows the two sizes of tins.

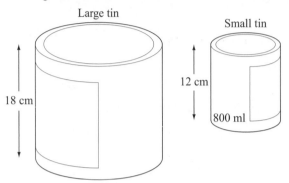

How many small tins can he fill from one large tin?

PS 8 All the lengths of the sides of a cube are increased by 10%.
    a   What is the percentage increase in the total surface area of the cube?
    b   What is the percentage increase in the volume of the cube?

AU 9 The length of a standard gift box is 10 cm.
The length of a large gift box is 15 cm.
The large box is similar to the standard box.

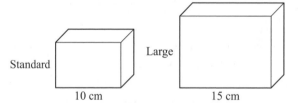

The volume of the standard box is 240 cm³.
Which of the following is the correct volume of the large box?
    a   360 cm³   **b**   540 cm³   **c**   720 cm³   **d**   810 cm³

HOMEWORK 7E

A*

 A bottling firm makes similar bottles in three different sizes: small, medium and large.
The volumes are:
    i small = 330 cm³   **ii** medium = 1000 cm³   **iii** large = 2000 cm³
    a   The medium bottle is 20 cm high. Find the heights of the other two bottles.
    b   The firm designs a label for the large bottle, and wants the labels on the other two
        bottles to be similar. If the area of the label on the large bottle is 100 cm², work out
        the area of the labels on the other two bottles.

2 It takes 1 kg of grass seed to cover a lawn that is 20 m long. How much seed will be needed to cover a similarly shaped lawn that is 10 m long?

3 Erin's model yacht has a mast that is 40 cm high. She dreams of having a real yacht with a mast that is 4 m high.

a The sail on her model yacht has an area of 600 cm². What is the area of the sail on the real yacht? Give your answer in m².

b The real yacht has a hull volume of 20 m³. What is the hull volume of her model yacht? Give your answer in cm³.

4 The ratio of the height of P to the height of Q is 5 : 4. The volume of P is 150 cm³. Calculate the volume of Q.

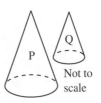

Not to scale

 5 Marie has two similar photographs.

The areas of the photographs are 200 cm² and 600 cm². Calculate the length *x* marked on the diagram.

 6 These two bottles of cola are similar in shape.

550 ml   850 ml

If the height of one of the bottles is 20 cm, calculate the two possible heights for the other bottle.

AU 7 The surface areas of two spheres are 108 cm² and 300 cm². The ratio of their volumes is given by which of the following?

a 3 : 5   **b** 9 : 25   **c** 27 : 125

Functional Maths Activity

Manufacturing plastic dice

Ms Newton runs a company which manufactures small items out of plastic. One item they make is a plastic dice with a side of 2 cm.

Ms Newton wants to make a larger dice. She needs to know what it will cost. This will depend on the amount of plastic required to make it.

1   She thinks that a dice with a side of 4 cm will require twice as much plastic to make as a dice with a side of 2 cm. Explain why this is not the case.
    How much more plastic will be required?

2   She wonders about making a dice with a side of 3 cm. How much more plastic will this use compared to a dice with a side of 2 cm?
    How much less plastic will a dice with a side of 3 cm require, compared to a dice with a side of 4 cm?

3   What advice can you give Ms Newton about the size of a dice that requires twice as much plastic to make one that is 2 cm?

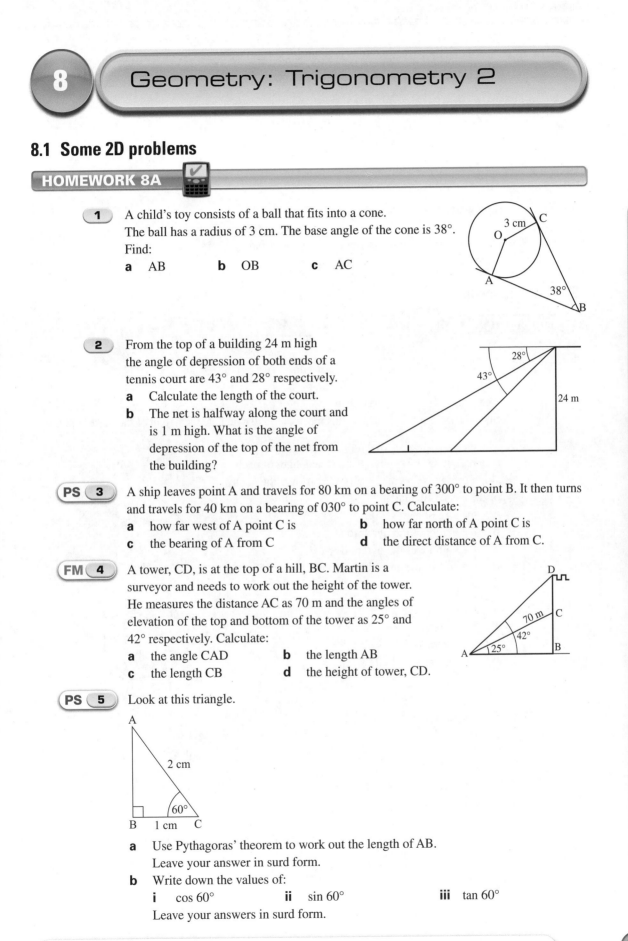

8 Geometry: Trigonometry 2

8.1 Some 2D problems

HOMEWORK 8A

1 A child's toy consists of a ball that fits into a cone.
The ball has a radius of 3 cm. The base angle of the cone is 38°.
Find:

a AB   **b** OB   **c** AC

2 From the top of a building 24 m high
the angle of depression of both ends of a
tennis court are 43° and 28° respectively.

a Calculate the length of the court.

b The net is halfway along the court and
is 1 m high. What is the angle of
depression of the top of the net from
the building?

PS 3 A ship leaves point A and travels for 80 km on a bearing of 300° to point B. It then turns
and travels for 40 km on a bearing of 030° to point C. Calculate:

a how far west of A point C is

b how far north of A point C is

c the bearing of A from C

d the direct distance of A from C.

FM 4 A tower, CD, is at the top of a hill, BC. Martin is a
surveyor and needs to work out the height of the tower.
He measures the distance AC as 70 m and the angles of
elevation of the top and bottom of the tower as 25° and
42° respectively. Calculate:

a the angle CAD   **b** the length AB

c the length CB   **d** the height of tower, CD.

PS 5 Look at this triangle.

a Use Pythagoras' theorem to work out the length of AB.
Leave your answer in surd form.

b Write down the values of:

i cos 60°   **ii** sin 60°   **iii** tan 60°

Leave your answers in surd form.

6 In the diagram, triangle ACD is right-angled and triangle ABC is isosceles.

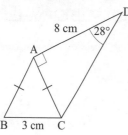

Calculate the size of angle ABC.

PS 7 A regular pentagon is inscribed in a circle of radius 5 cm.
Calculate the length of one of its sides.

8.2 Some 3D problems

HOMEWORK 8B ✔

FM 1 A TV mast XY is 3 km due west of village A.
Village B is 2 km due south of village A.
The angle of elevation from B to the top of the mast is 6°.
Show how a company can use this information to calculate the height of the mast in metres.

2 The diagram shows a pyramid. The base is a
square ABCD, 16 cm by 16 cm. The length of each
sloping edge is 25 cm. The apex, V, is over the centre
of the base. Calculate:

a the size of angle VAC

b the height of the pyramid

c the volume of the pyramid

d the size of the angle between the face VAD
and the base ABCD.

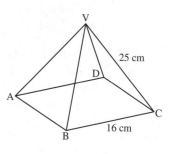

3 In the given cuboid, find:

a angle AGE

b angle BMA.

(M is the midpoint of GH.)

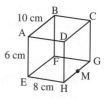

4 The diagram shows a wedge. Find:

a CD

b angle CAD

c angle CAE.

M is the midpoint of AB. Calculate:

d distance DM.

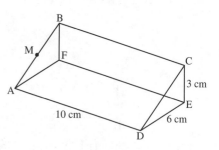

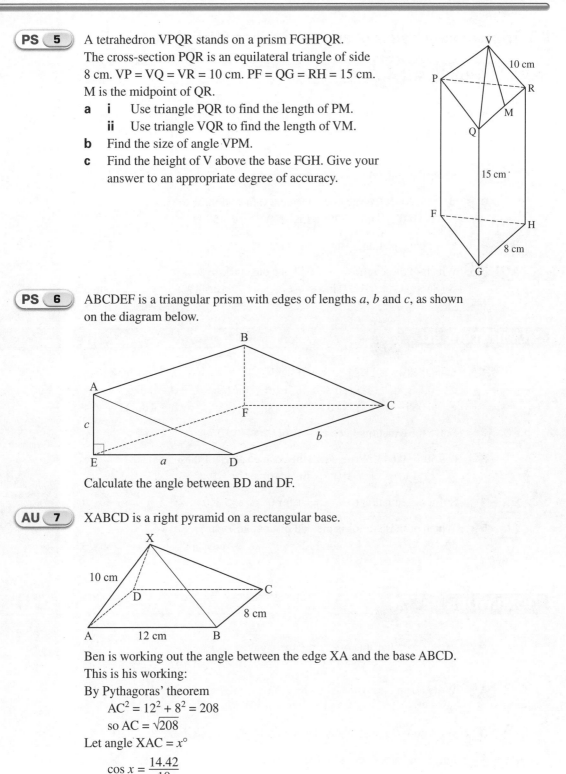

PS **5** A tetrahedron VPQR stands on a prism FGHPQR.
The cross-section PQR is an equilateral triangle of side
8 cm. VP = VQ = VR = 10 cm. PF = QG = RH = 15 cm.
M is the midpoint of QR.
- **a** **i** Use triangle PQR to find the length of PM.
  - **ii** Use triangle VQR to find the length of VM.
- **b** Find the size of angle VPM.
- **c** Find the height of V above the base FGH. Give your
  answer to an appropriate degree of accuracy.

PS **6** ABCDEF is a triangular prism with edges of lengths a, b and c, as shown
on the diagram below.

Calculate the angle between BD and DF.

AU **7** XABCD is a right pyramid on a rectangular base.

Ben is working out the angle between the edge XA and the base ABCD.
This is his working:
By Pythagoras' theorem
$$AC^2 = 12^2 + 8^2 = 208$$
$$so\ AC = \sqrt{208}$$
Let angle XAC = $x°$
$$\cos x = \frac{14.42}{10}$$

$$so\ x = \cos^{-1}\frac{14.42}{10}$$

Ben gets an error message on his calculator when he tries to work this out.
Explain where Ben has made the error and write out a correct solution to find
the value of x.

8.3 Trigonometric ratios of angles between 90° and 360°

HOMEWORK 8C

1 State the two angles between 0° and 360° for each of these sine values.
   a 0.4 **b** 0.45 **c** 0.65 **d** 0.27
   e 0.453 **f** −0.4 **g** −0.15 **h** −0.52

PS 2 Solve the equation $2 \sin x = 1$ for $0° \leqslant x \leqslant 360°$.

3 $\sin 40° = 0.643$. Write down the sine values of these angles.
   i 140° **ii** 320° **iii** 400° **iv** 580°

PS 4 Solve the equation $3 \sin x = -2$ for $0° \leqslant x \leqslant 360°$.

AU 5 Which of these ratios is the odd one out and why?
   sin 36°   sin 78°   sin 119°   sin 320°

HOMEWORK 8D

1 State the two angles between 0° and 360° for each of these cosine values.
   a 0.7 **b** 0.38 **c** 0.617 **d** 0.376
   e 0.085 **f** −0.6 **g** −0.45 **h** −0.223

PS 2 Solve the equation $3 \cos x = -1$ for $0° \leqslant x \leqslant 360°$.

3 $\cos 50° = 0.643$. Write down the cosine values of these angles.
   i 130° **ii** 310° **iii** 410° **iv** 590°

PS 4 Solve the equation $6 \cos x = -1$ for $0° \leqslant x \leqslant 360°$.

AU 5 Which of these ratios is the odd one out and why?
   cos 68°   cos 112°   cos 248°   cos 338°

HOMEWORK 8E

1 Write down the sine of each of these angles.
   a 27° **b** 153° **c** 207° **d** 333°

2 Write down the cosine of each of these angles.
   a 69° **b** 111° **c** 249° **d** 291°

3 What do you notice about the answers to Questions **1** and **2**?

AU 4 Find four values between 0° and 360° such that:
   a $\sin x = \pm 0.4$ **b** $\cos x = \pm 0.5$

PS 5 Solve: **a** $\sin x + 1 = 2$ for $0° \leqslant x \leqslant 360°$ **b** $2 + 3 \cos x = 1$ for $0° \leqslant x \leqslant 360°$

PS 6 Find two values of x between 0° and 360° such that $\sin x = \cos 320°$.

HOMEWORK 8F

1 State the angles between 0° and 360° for each of these tangent values.
  a 0.528   **b** 0.8   **c** 1.35   **d** 3.24
  e –2.55   **f** –0.158   **g** –0.786   **h** –1.999

AU 2 tan 64° = 2.05. Write down the tangent values of these angles.
  i 116°   **ii** 296°   **iii** 424°   **iv** 604°

8.4 Solving any triangle

HOMEWORK 8G

1 Find the length x in each of these triangles.

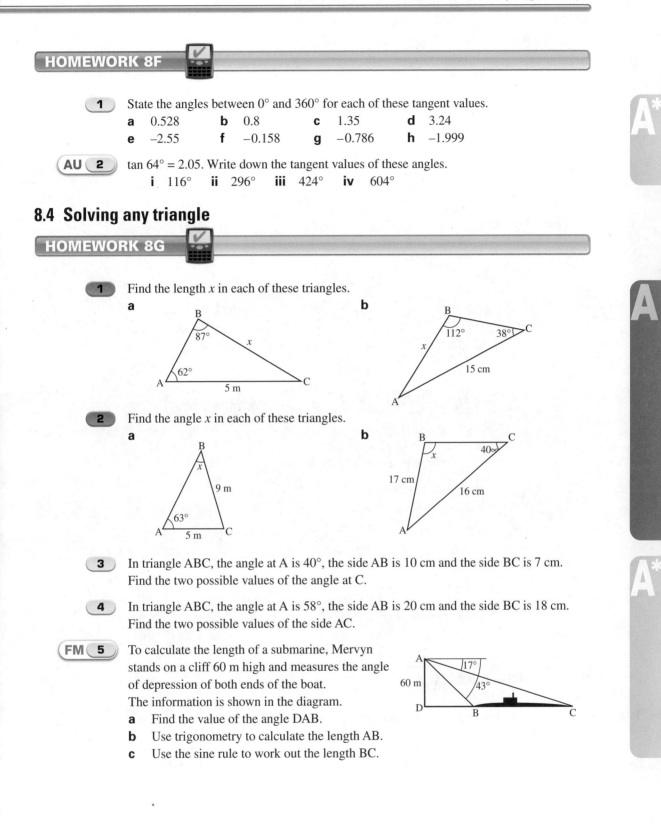

2 Find the angle x in each of these triangles.

3 In triangle ABC, the angle at A is 40°, the side AB is 10 cm and the side BC is 7 cm. Find the two possible values of the angle at C.

4 In triangle ABC, the angle at A is 58°, the side AB is 20 cm and the side BC is 18 cm. Find the two possible values of the side AC.

FM 5 To calculate the length of a submarine, Mervyn stands on a cliff 60 m high and measures the angle of depression of both ends of the boat. The information is shown in the diagram.
  a Find the value of the angle DAB.
  b Use trigonometry to calculate the length AB.
  c Use the sine rule to work out the length BC.

A*

FM 6 Use the information on this sketch to help
the land surveyor calculate the width, *w*,
of the river.

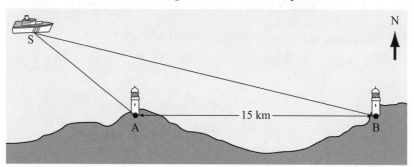

PS FM 7 A surveyor wishes to measure the height of a chimney. Measuring the angle of elevation,
she finds that the angle increases from 28° to 37° after walking 30 m towards the
chimney. What is the height of the chimney?

PS 8 Ship S and two lighthouses A and B are shown in the diagram below.
A is due west of B and the two lighthouses are 15 km apart.

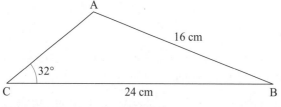

The bearing of the ship from lighthouse A is 330° and the bearing of the ship from
lighthouse B is 290°.
How far is the ship from lighthouse B?

9 Triangle ABC has an obtuse angle at A.

Calculate the size of angle BAC.

HOMEWORK 8H

A

1 Find the length *x* in each of these triangles.

a

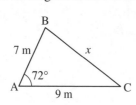

b

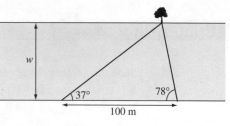

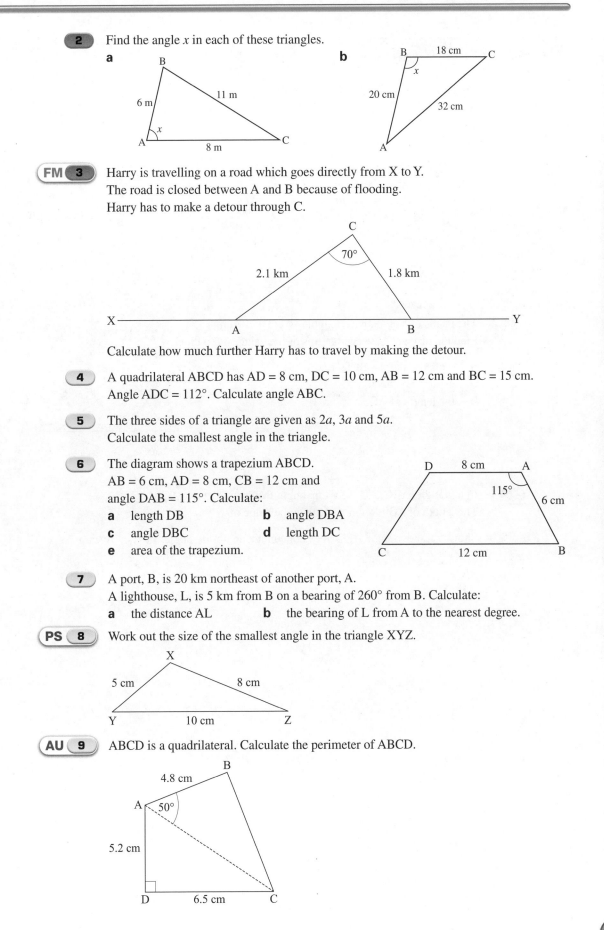

2 Find the angle *x* in each of these triangles.

a

B
6 m
11 m
x
A   8 m   C

b

B   18 cm   C
x
20 cm
32 cm
A

FM 3 Harry is travelling on a road which goes directly from X to Y.
The road is closed between A and B because of flooding.
Harry has to make a detour through C.

C
70°
2.1 km   1.8 km
X —————————————— Y
A   B

Calculate how much further Harry has to travel by making the detour.

4 A quadrilateral ABCD has AD = 8 cm, DC = 10 cm, AB = 12 cm and BC = 15 cm.
Angle ADC = 112°. Calculate angle ABC.

5 The three sides of a triangle are given as 2*a*, 3*a* and 5*a*.
Calculate the smallest angle in the triangle.

6 The diagram shows a trapezium ABCD.
AB = 6 cm, AD = 8 cm, CB = 12 cm and
angle DAB = 115°. Calculate:
a length DB   **b** angle DBA
c angle DBC   **d** length DC
e area of the trapezium.

D   8 cm   A
115°
6 cm
C   12 cm   B

7 A port, B, is 20 km northeast of another port, A.
A lighthouse, L, is 5 km from B on a bearing of 260° from B. Calculate:
a the distance AL   **b** the bearing of L from A to the nearest degree.

PS 8 Work out the size of the smallest angle in the triangle XYZ.

X
5 cm   8 cm
Y   10 cm   Z

AU 9 ABCD is a quadrilateral. Calculate the perimeter of ABCD.

B
4.8 cm
A   50°
5.2 cm
D   6.5 cm   C

HOMEWORK 8I

A

1 Find the length or angle x in each of these triangles.

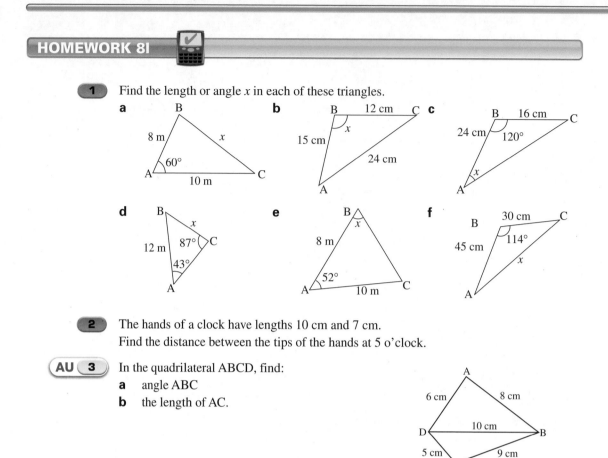

a B, 8 m, x, A, 60°, 10 m, C

b B, 12 cm, C, x, 15 cm, 24 cm, A

c B, 16 cm, C, 24 cm, 120°, A, x

d B, x, 12 m, 87°, C, 43°, A

e B, x, 8 m, A, 52°, 10 m, C

f B, 30 cm, C, 45 cm, 114°, A, x

2 The hands of a clock have lengths 10 cm and 7 cm.
Find the distance between the tips of the hands at 5 o'clock.

A*

AU 3 In the quadrilateral ABCD, find:
  a angle ABC
  b the length of AC.

A, 6 cm, 8 cm, D, 10 cm, B, 5 cm, 9 cm, C

AU 4 In a triangle, ABC, AC = 7.6 cm, angle BAC = 35°, angle ACB = 65°.
The length of AB is x cm. Calculate the value of x.

PS 5 Show that the triangle ABC does not contain an obtuse angle.

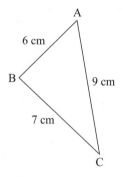

A, 6 cm, B, 9 cm, 7 cm, C

8.5 Trigonometric ratios in surd form

HOMEWORK 8J

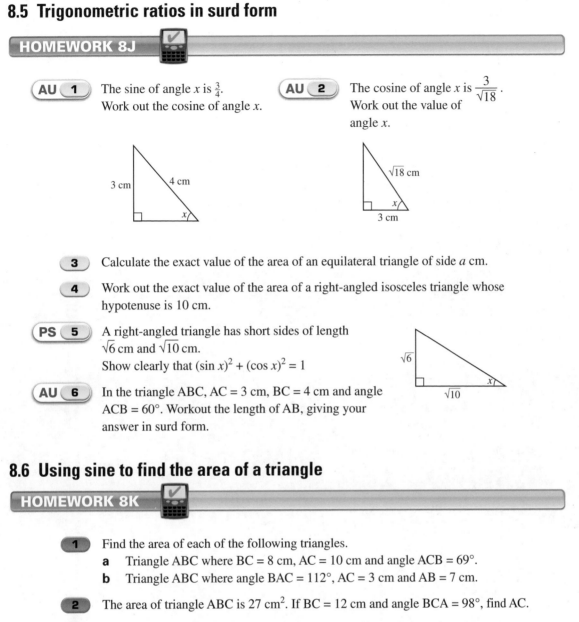

AU 1 The sine of angle x is $\frac{3}{4}$.
Work out the cosine of angle x.

AU 2 The cosine of angle x is $\frac{3}{\sqrt{18}}$.
Work out the value of angle x.

3 Calculate the exact value of the area of an equilateral triangle of side a cm.

4 Work out the exact value of the area of a right-angled isosceles triangle whose hypotenuse is 10 cm.

PS 5 A right-angled triangle has short sides of length $\sqrt{6}$ cm and $\sqrt{10}$ cm.
Show clearly that $(\sin x)^2 + (\cos x)^2 = 1$

AU 6 In the triangle ABC, AC = 3 cm, BC = 4 cm and angle ACB = 60°. Workout the length of AB, giving your answer in surd form.

8.6 Using sine to find the area of a triangle

HOMEWORK 8K

1 Find the area of each of the following triangles.
   a Triangle ABC where BC = 8 cm, AC = 10 cm and angle ACB = 69°.
   b Triangle ABC where angle BAC = 112°, AC = 3 cm and AB = 7 cm.

2 The area of triangle ABC is 27 cm². If BC = 12 cm and angle BCA = 98°, find AC.

3 In a quadrilateral ABCD, DC = 3 cm, BD = 8 cm, angle BAD = 43°, angle ABD = 52° and angle BDC = 72°. Calculate the area of the quadrilateral.

4 The area of triangle LMN is 85 cm², LM = 10 cm and MN = 25 cm. Calculate:
   a angle LMN   b angle MNL.

5 A signwriter wants to use a triangular shaped board with sides 30 cm, 40 cm and 60 cm. Help him to find its area.

AU 6 In the triangle ABC, angle B is **obtuse**, ∠BAC = 32°, AC = 10 cm, BC = 6 cm. Calculate the area of the triangle ABC.

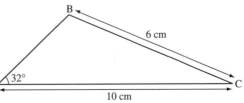

A*

FM **7** The diagram shows a sketch of the shape of a farmer's orchard.

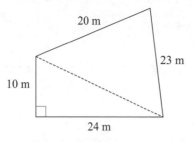

20 m

23 m

10 m

24 m

a Help the farmer to calculate the area of the orchard.
Give your answer to an appropriate degree of accuracy.

b For each 5 m² the farmer will plant a tree. How many trees can he plant in the orchard?

PS **8** ABCD is a parallelogram.
AB = a and BC = b.
Angle ABC = θ.

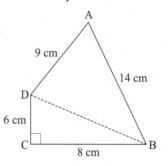

A   D

a

θ

B b C

Prove that the area, A, of the parallelogram is given by the formula:
A = $ab \sin \theta$.

AU **9** ABCD is a quadrilateral.

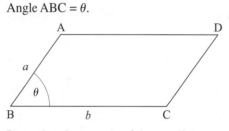

A

9 cm

14 cm

D

6 cm

C

8 cm   B

Work out the area of the quadrilateral.
Give your answer to an appropriate degree of accuracy.

Problem-solving Activity

Trigonometry and shapes

A formula for the area of a triangle with sides a, b and c is:

Area $= \sqrt{s(s-a)(s-b)(s-c)}$

where $s = \frac{1}{2}(a + b + c)$

1 Use this formula to find the area of a right-angled triangle with integer sides and show that it gives the correct answer.

2 Use the formula to find the area of an equilateral triangle. Find the area of this triangle by a different method and show that the formula gives the correct answer.

3 A piece of land is roughly triangular in shape. The sides of the triangle are 18 metres, 22 metres and 24 metres.
   What is the area of the piece of land?

4 A field is a quadrilateral in shape. You have been asked to find the area of the field. You have a long measuring tape to help you.
   What measurements would you take and how would you use them to find the area of the field?

9.1 Uses of graphs

B

1 This graph illustrates the charges made by an electricity company.
   a Calculate the standing charge. This is the amount paid before any electricity is used.
   b What is the gradient of the line?
   c From your answers to **a** and **b** write down the rule to calculate the total charge for electricity.

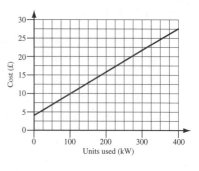

2 This graph illustrates the charges made by a gas company.
   a Calculate the standing charge.
   b What is the gradient of the line?
   c From your answers to **a** and **b** write down the rule to calculate the total charge for gas.

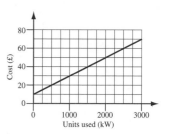

3 This graph illustrates the charges made by a phone company.
   a Calculate the standing charge.
   b What is the gradient of the line?
   c From your answers to **a** and **b** write down the rule to calculate the total charge for using this phone company.

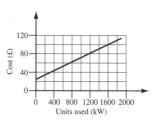

PS 4 The graph shows 3 line segments.

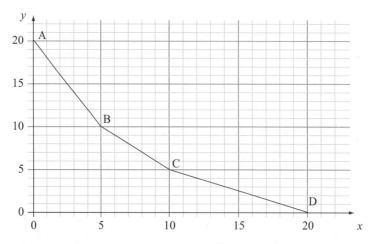

a Work out the equations of the lines through:
   i   A and B   **ii**   B and C   **iii**   C and D.

b Work out the area of the region between the line shown and AD.

HOMEWORK 9B

By drawing their graphs, find the solution of each of these pairs of simultaneous equations.

1 $x + 4y = 1$
    $x - y = 6$

2 $y = 2x + 1$
    $3x + 2y = 23$

3 $y = 2x + 5$
    $y = x + 4$

4 $y = x$
    $x - y = 4$

5 $y + 10 = 2x$
    $5x + y = 18$

6 $y = 5x - 1$
    $y = 3x + 2$

7 $y = x + 11$
    $x + y = 5$

8 $y - 3x = 8$
    $y = x + 6$

9 $y = -x$
    $y = 4x + 15$

10 $3x + 2y = 2$
    $y = -2x$

11 $y = 3x - 4$
    $y + x = 6$

12 $y = 3x - 12$
    $x + y = 2$

AU 13   Two coffees and three cakes cost £7.00.

Two coffees and one cake cost £4.00.

Using x to represent the price of a coffee and y to represent the cost of a cake, set up a pair of simultaneous equations.

Using a set of axes with the x-axis from 0 to 4 and the y-axis from 0 to 4, draw the graphs of the two equations.

Use the graphs to write down the cost of a coffee and the cost of a cake.

PS 14   The graph shows four lines:

P: $y = -x$ Q: $y = 2x + 6$ R: $y = x - 2$ S: $y = -\frac{2}{3}x - 2$

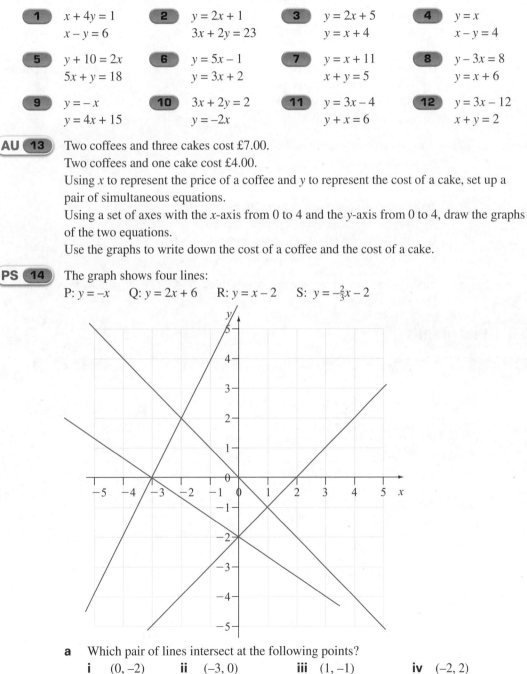

a Which pair of lines intersect at the following points?
   i   (0, −2)   **ii**   (−3, 0)   **iii**   (1, −1)   **iv**   (−2, 2)

b Solve the simultaneous equations given by P and S to find the exact solution.

9.2 Parallel and perpendicular lines

HOMEWORK 9C

FM 1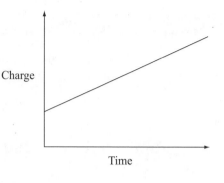

Two taxi companies, Acabs and BeeCabs, use the same graph to charge for journeys.
A sketch of this graph is shown here.

a Acabs decides to reduce its basic charge by 50% but to maintain the same charge per kilometre.
Sketch the new graph on a copy of the original graph.

b BeeCabs decide to have no basic charge but to double the charge per kilometre.
Sketch the new graph for BeeCabs on a copy of the original graph.

2 Write down the equation of the line parallel to each of the following lines and which passes through point (0, 1).

a $y = 2x - 3$ **b** $y = -4x + 3$ **c** $y = \frac{1}{2}x - 5$ **d** $y = -\frac{1}{4}x - 3$

3 Write down the equations of these lines.

a parallel to $y = 3x - 2$ and passes through (0, 4)

b parallel to $y = \frac{1}{4}x + 3$ and passes through (0, −1)

c parallel to $y = -x + 3$ and passes through (0, 2)

4 Find the equation of the line that passes through (4, 7) and is parallel to AB, where A is (1, 4) and B is (5, 2).

9.3 Other graphs

HOMEWORK 9D

1 **a** Complete the table to draw the graph of $y = x^3 + 1$ for $-3 \leqslant x \leqslant 3$

x	−3	−2	−1	0	1	2	3
$y = x^3 + 1$	−26			1			28

b Use your graph to find the y-value for an x-value of 1.2.

2 **a** Complete the table to draw the graph of $y = x^3 + 2x$ for $-2 \leqslant x \leqslant 3$

x	−2	−1	0	1	2	3
$y = x^3 + 2x$	−12		0		12	

b Use your graph to find the y-value for an x-value of 2.5.

3 **a** Complete the table to draw the graph of $y = \dfrac{12}{x}$ for $-12 \leqslant x \leqslant 12$

x	−12	−6	−4	−3	−2	−1	1	2	3	4	6	12
$y = \dfrac{12}{x}$	−1			−4					4			1

b Use your graph to find:

i the y-value when $x = 1.5$ **ii** the x-value when $y = 5.5$.

4   **a**   Complete the table to draw the graph of $y = \dfrac{50}{x}$ for $0 \leqslant x \leqslant 50$

x	1	2	5	10	25	50
$y = \dfrac{50}{x}$						

      b   On the same axes, draw the line $y = x + 30$

      c   Use your graph to find the x-value of the point where the graphs cross.

5   **a**   Complete the table below for $y = 2^x$ for values of x from -3 to $+4$. (Values are rounded to 2 dp.)

x	-3	-2	-1	0	1	2	3	4
$y = 2^x$	0.1	0.3			2	4		

      b   Plot the graph of $y = 2^x$ for $-3 \leqslant x \leqslant 4$ (Take y-axis from 0 to 20)

      c   Use your graph to estimate the value of y when $x = 2.5$

      d   Use your graph to estimate the value of x when $y = 0.75$

AU 6   A curve of the form $y = ab^x$ passes through the points $(0, 3)$ and $(2, 48)$. Work out the values of a and b.

9.4 The circular function graphs

HOMEWORK 9E

1   Given that $\sin 55° = 0.819$, find another angle between $0°$ and $360°$ that also has a sine of 0.819.

2   Given that $\sin 225° = -0.707$, find another angle between $0°$ and $360°$ that also has a sine of -0.707.

3   Given that $\cos 27° = 0.891$, find another angle between $0°$ and $360°$ that also has a cosine of 0.891.

4   Given that $\cos 123° = -0.545$, find another angle between $0°$ and $360°$ that also has a cosine of -0.545.

5   Given that $\sin 60° = 0.866$, find two angles between $0°$ and $360°$ that have a sine of -0.866.

6   Given that $\cos 30° = 0.866$, find two angles between $0°$ and $360°$ that have a cosine of -0.866.

7   Given that $\cos 38° = 0.788$:

      a   write down an angle between $0°$ and $360°$ that has a sine of 0.788

      b   find two angles between $0°$ and $360°$ that have a cosine of -0.788

      c   find two angles between $0°$ and $360°$ that have a sine of -0.788.

PS 8   **a**   Choose an obtuse angle a. Write down the values of:

           i $\sin a$ **ii** $\sin (180° - a)$

      b   Repeat with another acute angle b.

      c   Write down a rule connecting the sine of an obtuse angle x and the sine of the supplementary angle (i.e. the difference with $180°$).

      d   Find a similar rule for the cosine of x and the cosine of its supplementary angle.

AU 9 A formula used to work out the angle of a triangle is $\cos A \dfrac{b^2 + c^2 - a^2}{2bc}$, where a, b and c

are the sides of the triangle and angle A is the angle opposite side a.

Bill uses the formula to work out the angle A in this triangle, where $a = 18$, $b = 7$ and $c = 8$.

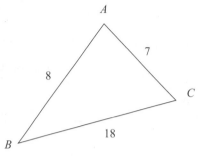

a Work out the value of $\cos A$ for these values.

b Explain why Bill cannot find a value for A.

9.5 Solving equations, one linear and one non-linear, with graphs

HOMEWORK 9F

Find the approximate or exact solutions to the following pairs of simultaneous equations using graphical methods.

The sizes of the axes needed are given in brackets.

1 $y = x^2 + 5x - 3$ and $y = x$ $(-10 \leqslant x \leqslant 5, -10 \leqslant y \leqslant 5)$

2 $x^2 + y^2 = 25$ and $x + y = 2$ $(-6 \leqslant x \leqslant 6, -6 \leqslant y \leqslant 6)$

3 $y = x^2 - 3x + 2$ and $y = x + 2$ $(-5 \leqslant x \leqslant 5, -5 \leqslant y \leqslant 5)$

4 $y = x^2 - 5$ and $y = x + 3$ $(-5 \leqslant x \leqslant 5, -6 \leqslant y \leqslant 10)$

PS 5 **a** $y = x^2 + 2x - 1$ and $y = 4x - 2$ $(-5 \leqslant x \leqslant 5, -5 \leqslant y \leqslant 10)$

     b What is special about the intersection of these two graphs?

     c Show that $4x - 2 = x^2 + 2x - 1$ can be rearranged to: $x^2 - 2x + 1 = 0$

     d Factorise and solve: $x^2 - 2x + 1 = 0$

     e Explain how the solution in **d** relates to the intersection of the graphs.

AU 6 **a** $y = x^2 + 3x + 5$ and $y = 2x - 1$ $(-5 \leqslant x \leqslant 5, -5 \leqslant y \leqslant 8)$

     b What is special about the intersection of these two graphs?

     c Rearrange $2x - 1 = x^2 + 3x + 5$ into the general quadratic form: $ax^2 + bx + c = 0$

     d Work out the discriminant $b^2 - 4ac$ for the quadratic in **c**.

     e Explain how the value of the discriminant relates to the intersection of the graphs.

9.6 Solving equations by the method of intersection

HOMEWORK 9G

1 Below is the graph of: $y = x^2 - 2x - 4$

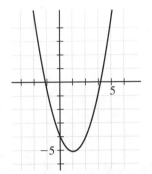

Use this graph to solve:
a $x^2 - 2x - 4 = 0$
b $x^2 - 2x - 4 = 4$
c $x^2 - 2x - 3 = 0$

PS 2 Below are the graphs of: $y = x^2 - 3x + 1$, $y = x - 1$ and $y + x = 2$

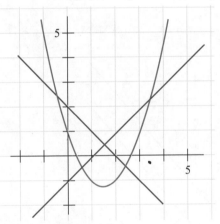

Use these graphs to solve:
a $x^2 - 3x + 1 = 0$
b $2x - 3 = 0$
c $x^2 - 3x - 1 = 0$
d $x^2 - 4x + 2 = 0$
e $x^2 - 2x - 1 = 0$

3 Draw the graph of: $y = x^3 - 2x + 3$
a Use the graph to solve: **i** $x^3 - 2x + 3 = 0$ **ii** $x^3 - 2x = 0$
b Draw a straight-line graph to solve: $x^3 - 3x + 2 = 0$
Draw this line and solve: $x^3 - 3x + 2 = 0$

4 The graph of $y = x^3 - 4x - 1$ is shown on the right.

    a Use the graph to solve:

      i $x^3 - 4x - 1 = 0$

      ii $x^3 - 4x + 2 = 0$

    b By drawing an appropriate straight line, solve the equation: $x^3 - 5x - 1 = 0$

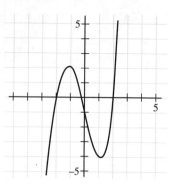

5 The graph of $y = x^3 - 4x$ is shown.

    a Use the graph to find the two positive solutions to: $x^3 - 4x = -2$

    b By drawing an appropriate straight line, use the graph to solve: $x^3 - 3x + 1 = 0$

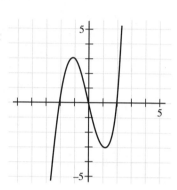

PS 6 The graph shows the lines A: $y = x^2 + 5x - 3$; B: $y = x$; C: $y = x + 3$; D: $y + x = 2$ and E: $y = -x$

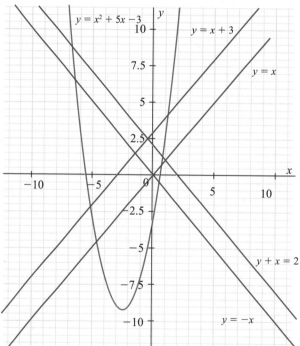

    a Which pair of lines has a common solution of $(-1.5, 1.5)$?

    b Which pair of lines has the approximate solutions $(1, 1)$ and $(-6.8, 8.8)$?

    c What quadratic equation has an approximate solution of $(-5.2, -2.2)$ and $(1.2, 4.2)$?

    d The minimum point of the graph $y = x^2 + 5x - 3$ is at $(-2.5, -9.25)$.

      Write down the minimum point of the graph: $y = x^2 + 5x - 8$

Problem-solving Activity

Granny's will

Granny has two nephews, Alf and Bert. She writes a will leaving Alf £1000 in the first year after her death, £2000 in the second year after her death, £3000 the next year and so on for 20 years. She leaves Bert £1 the first year, £2 the second year, £4 the next year and so on.

a Show that the formula $500n(n + 1)$ gives the total amount that Alf gets after n years.

b Show that the formula $2^n - 1$ gives the total amount that Bert gets after n years.

c Complete the table for the total amount of money that Bert gets.

Year	2	4	6	8	10	12	14	16	18	20
Total	3	15	63							

d Draw a graph of both nephews' total over 20 years. Take the x-axis from 0 to 20 years and the y-axis from £0 to £110 000.

e Which nephew gets the better deal?

10 Statistics: Data distributions

10.1 Cumulative frequency diagrams

1 An army squad were all sent on a one mile run. Their coach recorded the times they actually took. This table shows the results.

Time (seconds)	Number of players
$200 < x \leq 240$	3
$240 < x \leq 260$	7
$260 < x \leq 280$	12
$280 < x \leq 300$	23
$300 < x \leq 320$	7
$320 < x \leq 340$	5
$340 < x \leq 360$	5

    a Copy the table and complete a cumulative frequency column.

    b Draw a cumulative frequency diagram.

    c Use your diagram to estimate the median time and the interquartile range.

2 A company had 360 web pages. They recorded how many times they were visited on one day.

Number of visits	Number of pages
$0 < x \leq 50$	6
$50 < x \leq 100$	9
$100 < x \leq 150$	15
$150 < x \leq 200$	25
$200 < x \leq 250$	31
$250 < x \leq 300$	37
$300 < x \leq 350$	32
$350 < x \leq 400$	17
$400 < x \leq 450$	5

    a Copy the table and complete a cumulative frequency column.

    b Draw a cumulative frequency diagram.

    c Use your diagram to estimate the median use of the web pages and the interquartile range.

    d Pages with less than 60 visitors are going to be rewritten. About how many pages would need to be rewritten?

FM 3 For the mock exams, two classes were given two papers – Paper 1 and Paper 2. The results were summarised in the cumulative frequency graphs opposite.

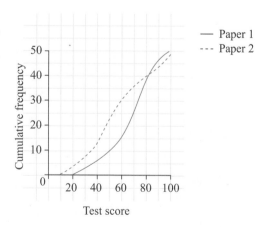

    a What is the median score for each paper?

    b What is the interquartile range for each paper?

    c Which is the harder paper? Explain how you know.

    d The teachers wanted 90% of the students to pass each paper and 15% of the students to get top marks in each paper. What marks for each paper give:

      i a pass

      ii the top grade.

FM Functional Maths   **AU** (AO2) Assessing Understanding   **PS** (AO3) Problem Solving

PS **4** The lengths of time, in minutes, that it took for the Ambulance Service to get an ambulance to a patient and then to a hospital, were recorded. A cumulative frequency diagram of this data is shown opposite.
Calculate the estimated mean length of time it took the Ambulance Service to get to a patient and then to a hospital.

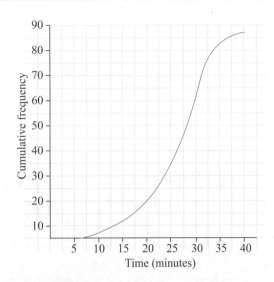

AU **5** Johnny was given a cumulative frequency diagram showing the number of students gaining marks in a spelling test. He was told the top 15% were given the top grade. How would you find the marks needed to gain this top award?

10.2 Box plots

HOMEWORK 10B

1 The box plot below shows the number of peas in pods grown by a prize-winning gardener.

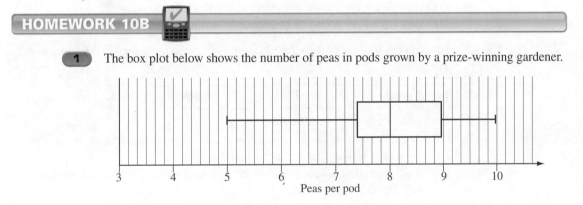

Peas per pod

A young gardener also grew some peas. These are the results for the number of peas per pod. Smallest number 3, Lower quartile 4.75, Median 5.5, Upper quartile 6.25, Highest number 9.

a Copy the diagram and draw a box plot for the young gardener.
b Comment on the differences between the two distributions.

2 The box plot shows the monthly salaries of the men in a computer firm.

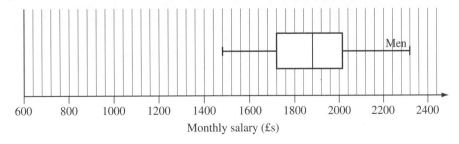

Monthly salary (£s)

The data for the women in the company is Smallest 600, Lower quartile 1300, Median 1600, Upper Quartile 2000, Largest 2400.

a Copy the diagram and draw a box plot for the women's salaries.
b Comment on the differences between the two distributions.

B

3 The box plots for the hours of life of two brands of batteries are shown below.

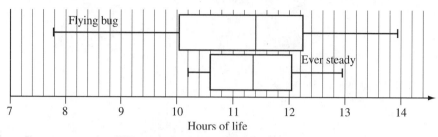

Hours of life

a Comment on the differences in the two distributions.

b Mushtaq wants to get some batteries for his palm top. Which brand would you recommend and why?

4 The following table shows some data on the times of telephone calls to two operators at a mobile phone helpline.

	Lowest time	Lowest Quartile	Median time	Upper Quartile	Highest time
Jack	1 m 10 s	2 m 20 s	3 m 30 s	4 m 50 s	7 m 10 s
Jill	40 s	2 m 20 s	5 m 10 s	7 m 30 s	10 m 45 s

a Draw box plots to compare both sets of data.

b Comment on the differences between the distributions.

c The company has to get rid of 1 operator. Who should go and why?

5 A school entered 80 pupils for an examination. The results are shown in the table.

Mark, x	$0 < x \leq 20$	$20 < x \leq 40$	$40 < x \leq 60$	$60 < x \leq 80$	$80 < x \leq 100$
Number of pupils	2	14	28	26	10

a Calculate an estimate of the mean.

b Complete a cumulative frequency table and draw a cumulative frequency diagram.

c **i** Use your graph to estimate the median mark.

  ii 12 of these pupils were given a grade A. Use your graph to estimate the lowest mark for which grade A was given.

d Another school also entered 80 pupils for the same examination. Their results were Lowest mark 40, Lower quartile 50, Median 60, Upper quartile 70, Highest mark 80. Draw a box plot to show these results and use it to comment on the differences between the two schools' results.

FM 6 A dental practice had two doctors: Dr Ball and Dr Charlton.

The following box plots were created to illustrate the waiting times for their patients during November.

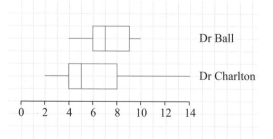

Gabriel was deciding which doctor to try and see. Which one would you advise he see and why?

PS 7 The box plots for a schools end of year science tests are shown below.

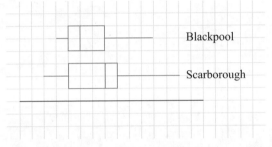

What is the difference between the means of the boys' and the girls' test results?

AU 8 Joy was given a diagram showing box plots for the daily sunshine in the seaside resorts of Scarborough and Blackpool for July, but no scale was shown. She was told to write a report on the differences between the sunshine in both resorts.

Invent a report that could be possible for her to make from box plots with no scales shown.

10.3 Histograms with bars of unequal width

HOMEWORK 10C

1 a The table shows the ages of 300 people at the cinema.

Age, x years	$0 \leqslant x < 20$	$20 \leqslant x < 30$	$30 \leqslant x < 50$
Frequency	110	115	75

Draw a histogram to show the data.

b Compare your histogram to the one you drew in Homework 3E Question 1a. What do you notice?

Age, x years	$20 \leqslant x < 30$	$30 \leqslant x < 40$	$40 \leqslant x < 50$	$50 \leqslant x < 60$	$60 \leqslant x < 70$
Frequency	35	120	130	50	15

Draw a histogram to show this data.

2 For the histogram on the right.
a write out the frequency diagram
b calculate an estimate of the mean of the distribution.

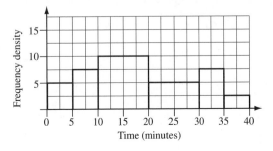

A

A*

3 The waiting times for customers at a supermarket checkout are shown in the table.
  a Draw a histogram of these waiting times.
  b Show an estimate of the median on your histogram. Show your working.

Waiting time (minutes)	Frequency
$0 \leqslant x < 1$	15
$1 \leqslant x < 3$	7
$3 \leqslant x < 4$	12
$4 \leqslant x < 5$	15
$5 \leqslant x < 10$	12

 4 Andrew was asked to create a histogram.
Explain to Andrew how he can find the height of each bar on the frequency density scale.

 5 The histogram shows the science test scores for students in a school.

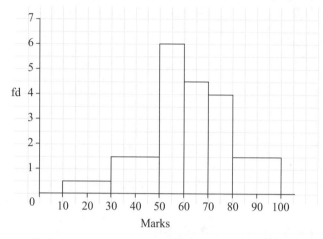

  a Estimate the median score.
  b Estimate the inter quartile range of the scores.
  c Find an estimate for the mean score.
  d What was the score needed for an A if 10% of the students gained an A?

PS 6 The distances that the members of a church travel to the church are shown in the histogram on the right. It is known that 22 members travel between 4 km and 6 km to church. What is the probability of choosing a member at random who travels more than 4 km to church?

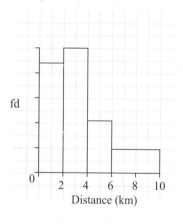

Functional Maths Activity

Buying cheeses

Marco is the head chef at Giorgio's restaurant. He regularly has to select and order a range of top quality cheeses to use in his recipes.

Here is a table showing the total weights of cheeses he has bought over a number of years.

| Cheeses | Weight in kilograms | | | | |
	2005	2006	2007	2008	2009
Wensleydale	26	34	33	35	38
Brie	17	13	16	21	22
Red Cheddar	13	17	21	22	26
White Cheshire	16	17	21	20	19
Red Leicester	13	12	16	22	23
Stilton	18	20	18	21	22
Edam	8	7	7	8	7

Marco wants to hand over responsibility for selecting and purchasing cheeses to a new buying manager. The new manager has asked for some information. Help Marco prepare the following:

1   Select the most appropriate statistical diagrams and measures to summarise the data given in the table.

2   Write a report about the cheeses used at Marco's restaurant over these five years.

11 Probability: Calculating probabilities

11.1 Tree diagrams

HOMEWORK 11A

1 A dice is thrown twice. Copy and complete the tree diagram on the right to show all the outcomes. Use your tree diagram to work out the probability of:

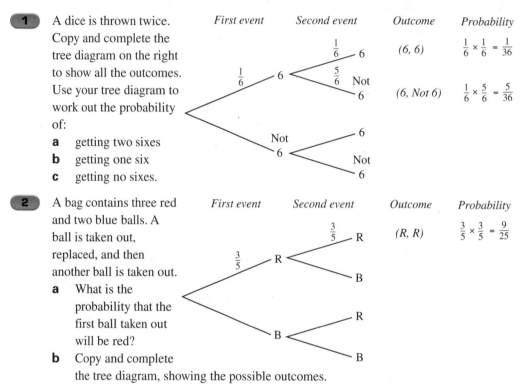

a getting two sixes
b getting one six
c getting no sixes.

2 A bag contains three red and two blue balls. A ball is taken out, replaced, and then another ball is taken out.

a What is the probability that the first ball taken out will be red?

b Copy and complete the tree diagram, showing the possible outcomes.

c Using the tree diagram, what is the probability of the following outcomes?
   i two red balls   ii exactly one red ball   iii I get at least one red ball.

3 A card is drawn from a pack of cards. It is replaced, the pack is shuffled and another card is drawn.

a What is the probability that either card was a Spade?
b What is the probability that either card was not a Spade?
c Draw a tree diagram to show the outcomes of two cards being drawn as described. Use the tree diagram to work out the probability that:
   i both cards will be Spades
   ii at least one of the cards will be a Spade.

4 A bag of sweets contains five chocolates and four toffees. I take two sweets out at random and eat them.

a What is the probability that the first sweet chosen is:
  i a chocolate **ii** a toffee?

b If the first sweet chosen is a chocolate:
  i how many sweets are left to choose from
  ii how many of them are chocolates?

c If the first sweet chosen is a toffee:
  i how many sweets are left to choose from
  ii how many of them are toffees?

d Copy and complete the tree diagram.

e Use the tree diagram to work out the probability that:
  i both sweets will be of the same type **ii** there is at least one chocolate chosen.

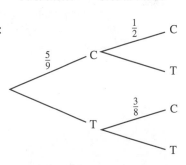

First choice   *Second choice*

FM 5 Thomas has to take a driving test which is in two parts. The first part is theoretical. He has a 0.4 chance of passing this. The second is practical. He has a 0.5 chance of passing this. Draw a tree diagram covering passing or failing the two parts of the test. What is the probability that he passes both parts?

6 Early every Sunday morning Carol goes out for a run. She has three pairs of shorts of which two are red and one is blue. She has five T-shirts of which three are red and two are blue. Because she doesn't want to disturb her sleeping family she gets dressed in the dark and picks a pair of shorts and a T-shirt at random.

a What is the probability that the shorts are blue?

b Copy and complete this tree diagram.

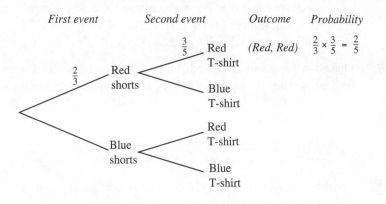

c What is the probability that Carol goes running in:
  i a matching pair of shorts and T-shirt
  ii a mismatch of shorts and T-shirt
  iii at least one red item?

7 Bob has a bag containing four blue balls, five red balls and one green ball. Sally has a bag containing two blue balls and three red balls. The balls are identical except for the colour. Bob chooses a ball at random from his bag; Sally chooses a ball at random from her bag.

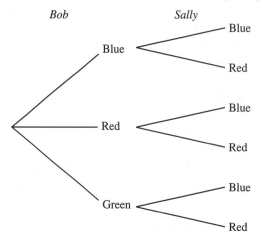

a On a copy of the tree diagram, write the probability of each of the events on the appropriate branch.

b Calculate the probability that both Bob and Sally will choose a blue ball.

c Calculate the probability that the ball chosen by Bob will be a different colour from the ball chosen by Sally.

PS 8 When playing the game 'Rushdown', you are dealt two cards. If you are dealt any two cards which are either 6 or 7 or 8, you have been dealt a 'Tango'.
What is the probability of being dealt a 'Tango'?
Give your answer to 3 decimal places.

AU 9 I have a drawer containing black, blue and brown socks. Explain how a tree diagram can help me find the probability of picking at random two socks of the same colour.

11.2 Independent events 1

HOMEWORK 11B

1 Ahmed throws a dice twice. The dice is biased so the probability of getting a six is $\frac{1}{4}$.
What is the probability that he gets:
a two sixes   **b** exactly one six?

2 Betty draws a card from a pack of cards, replaces it, shuffles the pack and then draws another card. What is the probability that the cards are:
a both Hearts   **b** a Heart and a Spade (in any order)?

3 Colin draws a card from a pack of cards, does not replace it and then draws another card. What is the probability that the cards are:
a both Hearts   **b** a Heart and a Spade (in any order)?

4 I throw a dice three times. What is the probability of getting a total score of 17 or 18?

5 A bag contains seven white beads and three black beads. I take out a bead at random, replace it and take out another bead. What is the probability that:
a both beads are black   **b** one bead is black and the other white (in any order)?

6 A bag contains seven white beads and three black beads. I take out a bead at random, do not replace it and take out another bead. What is the probability that:

a both beads are black **b** one bead is black and the other white (in any order)?

7 When I answer the telephone the call is never for me. Half the calls are for my daughter Janette. One-third of them are for my son Glen. The rest are for my wife Barbara.

a I answer the telephone twice this evening. Calculate the probability that:

i the first call will be for Barbara **ii** both calls will be for Barbara.

b The probability that both calls are for Janette is $\frac{1}{4}$. The probability that they are both for Glen is $\frac{1}{9}$. Calculate the probability that they are both for Janette or both for Glen.

FM 8 Anne regularly goes to London by train.

The probability of the train arriving in London late is 0.08.

The probability of the train being early is 0.02.

The probability of it raining in London is 0.3.

What is the probability of:

a Anne getting to London on time and it not raining

b Anne travelling to London three days in a row and it raining every day

c Anne travelling to London five days in a row and not being late at all?

PS 9 What is the probability of rolling the same number on a dice three times in a row?

AU 10 Give two events that could be described as 'independent'.

HOMEWORK 11C

1 Steve eats in the school canteen five days a week. The probability that there is turnip on the menu on any day is $\frac{1}{4}$.

a What is the probability that in five days there is no turnip on the menu?

b What is the probability that Steve gets turnip at least once in five days?

2 Three coins are thrown. What is the probability of getting:

a three tails **b** at least one head?

3 Adam is a talented all-round athlete. He is entered for the 100 m race, the javelin and the high jump. The probability that he wins each of these events is $\frac{4}{5}$, $\frac{3}{4}$ and $\frac{1}{2}$ respectively.

a What is the probability that he doesn't win any event?

b What is the probability that he wins at least one event?

4 A bag contains seven red and three blue balls. A ball is taken out and replaced. Another ball is taken out. What is the probability that:

a both balls are red **b** both balls are blue **c** at least one is red?

5 A bag contains seven red and three blue balls. A ball is taken out and not replaced. Another ball is taken out. What is the probability that:

a both balls are red **b** both balls are blue **c** at least one is red?

6 **a** A coin is thrown three times. What is the probability of:

i three heads **ii** no heads **iii** at least one head?

b A coin is thrown four times. What is the probability of:

i four heads **ii** no heads **iii** at least one head?

c A coin is thrown five times. What is the probability of:

i five heads **ii** no heads **iii** at least one head?

d A coin is thrown n times. What is the probability of:

i n heads **ii** no heads **iii** at least one head?

A

PS 7 The Batemore town show lasts for five days Wednesday to Sunday in a week during April. In this town, at this time of year, the probability of rain on any day during the show is 0.15.

    a What is the probability of:
        i a day with no rain during the show?
        ii two days in a row with no rain?
    b On how many days of the show could they expect:
        i rain?
        ii no rain at all?
    c What is the probability of getting rain on at least one of the days of the show?

PS 8 The probability of planting a banana tree in Bajora and it growing well is 0.7. Kiera plants 10 banana trees in Bajora. What is the probability that at least nine banana trees will grow well?

AU 9 In a class there are 30 students. You are told the probability of a student not being well on the exam day. How will you find the probability of at least one of the students not being well on the exam day?

10 A class has 12 boys and 14 girls. Three of the class are to be chosen at random to do a task for their teacher. What is the probability that:
    a all three are girls   **b** at least one is a boy?

11 From three boys and two girls, two children are to be chosen to present a bouquet of flowers to the Mayoress. The children are to be selected by drawing their names out of a hat. What is the probability that:
    a there will be a boy and a girl chosen
    b there will be two boys chosen
    c there will be at least one girl chosen?

11.3 Independent events 2

HOMEWORK 11D

1 A bag contains two black balls and five red balls. A ball is taken out and replaced. This is repeated three times. What is the probability that:
    a all three are black   **b** exactly two are black
    c exactly one is black   **d** none is black?

2 A bag contains two blue balls and five white balls. A ball is taken out but not replaced. This is repeated three times. What is the probability that:
    a all three are blue   **b** exactly two are blue
    c exactly one is blue   **d** none is blue?

3 A dice is thrown three times. What is the probability that:
    a four sixes are thrown   **b** no sixes are thrown   **c** exactly one six is thrown?

4 Ann is late for school with a probability of 0.7. Bob is late with a probability of 0.6. Cedrick is late with a probability of 0.3. On any particular day what is the probability of:
    a exactly one of them being late   **b** exactly two of them being late?

5 Daisy takes three AS exams in Mathematics. The probability that she will pass Pure 1 is 0.9. The probability that she will pass Statistics 1 is 0.65. The probability she will pass Discrete 1 is 0.95. What is the probability that she will pass:
    a all three modules   **b** exactly two modules   **c** at least two modules?

6 Six out of 10 cars in Britain run on petrol. The rest use diesel fuel. Three cars can be seen approaching in the distance.

    a What is the probability that the first one uses petrol?

    b What is the probability that exactly two of them use diesel fuel?

    c Explain why, if the first car uses petrol, the probability of the second car using petrol is still 6 out of 10.

7 Six T-shirts are hung out at random on a washing line. Three are red and three are blue. Using R and B for red and blue, write down all 20 possible combinations: for example, RRRBBB, RRBBBR and so on. What is the probability of:

    a three red shirts being next to each other

    b three blue shirts being next to each other?

8 In a class of 30 pupils, 21 have dark hair, seven have fair hair and two have red hair. Two pupils are chosen at random to collect homework.

    a What is the probabilty that they:

        i both have fair hair

        ii each have hair of a different colour?

    b If three pupils are chosen, what is the probability that exactly two have dark hair?

9 A firm is employing temporary workers. They call 30 for interview. It is found out that seven of the interviewees have a criminal record.

The firm hires three of the interviewees. What is the probability that:

    a all three have a criminal record

    b only one has a criminal record

    c at least two have a criminal record?

AU 10 Paul was playing a card game and was dealt three cards, all Aces. He thought the chance of him now being dealt another Ace was $\frac{1}{52}$.

Explain why he was wrong.

11.4 Conditional probability

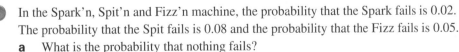

HOMEWORK 11E

1 Two tetrahedral (four-sided) dice, whose faces are numbered 1, 2, 3 and 4, are thrown together. The score on the dice is the uppermost face. Show by means of a sample space diagram that there are 16 outcomes with total scores from 2 to 8. What is the probability of:

    a a total score of 8   **b** at least one 3 on either dice?

2 Based on previous results the probability that Manchester United win is $\frac{2}{3}$, the probability they draw is $\frac{1}{4}$ and the probability they lose is $\frac{1}{12}$.

They play three matches. What is the probability that:

    a they win all three   **b** they win exactly two matches

    c they win at least one match?

3 In the Spark'n, Spit'n and Fizz'n machine, the probability that the Spark fails is 0.02. The probability that the Spit fails is 0.08 and the probability that the Fizz fails is 0.05.

    a What is the probability that nothing fails?

    b The machine will still work with one component out of action. What is the probability that it works?

A

A*

4 On average, Steve is late for school two days each week (of five days).
- **a** What is the probability that he is late on any one day?
- **b** In a week of five days, what is the probability that:
  - **i** he is late every day
  - **ii** he is late exactly once
  - **iii** he is never late
  - **iv** he is late at least once?

PS 5 The probability of me waking up early at the weekend on exactly one of those days (Saturday or Sunday) is 0.42.

What is the probability of me waking up early on any day?

6 A bag contains five black and three white balls. Three balls are taken out one at a time.
- **a** If the balls are put back each time, what is the probability of getting:
  - **i** three black balls
  - **ii** at least one black ball?
- **b** If the balls are not put back each time, what is the probability of getting:
  - **i** three black balls
  - **ii** at least one black ball?

7 Two cards are drawn one at a time from a pack of cards. The cards are replaced each time. What is the probability that at least one of them is a Heart?

8 Two cards are drawn from a pack of cards. The cards are not replaced each time. What is the probability that at least one of them is a Heart?

9 A box contains 50 batteries. It is known that 20 of them are dead. John needs two batteries for his calculator. He takes three out. What is the probability that:
- **a** all three of them are dead
- **b** at least one of them works?

10 From a box of 100 batteries, Janet takes out three batteries for her radio. The radio will work if two or three of the batteries are good. What is the probability that the radio will work?

11 Dan has six socks in a drawer, of which four are blue and two are black. He takes out two socks. What is the probability that:
- **a** both socks are blue
- **b** both socks are black
- **c** he gets a pair of socks
- **d** at least one of the socks is blue?

12 One in nine people is left-handed. Five people are in a room. What is the probability that:
- **a** all five are left-handed
- **b** all five are right-handed
- **c** at least one of them is left-handed?

PS 13 In a pack of cards, the Aces, the Kings, the Queens and the Jacks are all called 'colour cards'.

What is the probability of being dealt four 'colour cards' in a row from a normal deck of cards?

AU 14 A bag of jelly babies contains some yellow, some green and some orange jelly babies, all the same size. Trisha is asked to find the probability of taking out two jelly babies of the same colour.

Explain to Trisha how you would do this, explaining carefully the point where she is most likely to go wrong.

Functional Maths Activity

Social-club lottery

A social club runs a lottery, using the numbers 1 to 10. Three numbers are chosen each week.

1 Margaret chooses the numbers, 6, 7 and 8.
What is her probability of winning the lottery?

2 Richard says Margaret is never going to win with three consecutive numbers. He chooses the ages of his nieces and nephews which are 2, 5 and 7.
He says, 'I have a better chance than Margaret.'
Is he correct? Explain your answer.

3 There are 45 members in the social club. Every week they all choose three numbers and pay 20p.
If anyone's three numbers are selected, that person wins £5. There are no rollovers.
Throughout the year, the lottery is run 50 times.
How much profit would the social club expect to make from this venture in one year?

12.1 Rearranging formulae

HOMEWORK 12A

FM 1 A restaurant has a large oven that can cook up to 10 chickens at a time.
The restaurant uses the following formula for the length of time it takes to cook
n chickens:

$$T = 10n + 55$$

A large party is booked for a chicken dinner at 7 pm. They will need eight chickens
between them.

a It takes 15 minutes to get the chickens out of the oven and prepare them for serving.
At what time should the eight chickens go into the oven?

b The next day another large party is booked.

   i Rearrange the formula to make n the subject.

   ii The party is booked for 8 pm and the chef calculates she will need to put the
chickens in the oven at 5.50 pm.
How many chickens do the party need?

AU 2 Kern notices that the price of six coffees is 90 pence less than the price of nine teas.
Let the price of a coffee be x pence and the price of a tea be y pence.

a Express the cost of a tea, y, in terms of the price of a coffee, x.

b If the price of a coffee is £1.20, how much is a tea?

PS 3 Distance, speed and time are connected by the formula:
Distance = Speed ↔ Time
A delivery driver drove 90 miles at an average speed of 60 miles per hour.
On the return journey, he was held up at some road works for 30 minutes.
What was his average speed on the return journey?

4 $y = mx + c$ **a** Make c the subject. **b** Express x in terms of y, m and c.

5 $v = u - 10t$ **a** Make u the subject. **b** Express t in terms of v and u.

6 $T = 2x + 3y$ **a** Express x in terms of T and y. **b** Make y the subject.

7 $p = q^2$ Make q the subject.

8 $p = q^2 - 3$ Make q the subject.

9 $a = b^2 + c$ Make b the subject.

10 A rocket is fired vertically upwards with an initial velocity of u metres per second. After
t seconds the rocket's velocity, v metres per second, is given by the formula $v = u + gt$,
where g is a constant.

a Calculate v when $u = 120$, $g = -9.8$ and $t = 6$

b Rearrange the formula to express t in terms of v, u, and g.

c Calculate t when $u = 100$, $g = -9.8$ and $v = 17.8$

FM Functional Maths **AU** (AO2) Assessing Understanding **PS** (AO3) Problem Solving

12.2 Changing the subject of a formula

HOMEWORK 12B

In questions **1** to **5**, make the stated term the subject of each formula.

1 $4(x - 2y) = 3(2x - y)$ (x)

2 $p(a - b) = q(a + b)$ (a)

3 $A = 2ab^2 + ac$ (a)

4 $s(t + 1) = 2r + 3$ (r)

5 $st - r = 2r - 3t$ (t)

6 Make x the subject of these equations.

    a $ax = b - cx$
    b $x(a - b) = x + b$

    c $a - bx = dx - a$
    d $x(c - d) = c(d - x)$

7 **a** The perimeter of the shape on the right is given by the formula $P = 2\pi r + 4r$
    Make r the subject of the formula.

    b The area of the same shape is given by $A = \pi r^2 + 4r^2$
    Make r the subject of this formula.

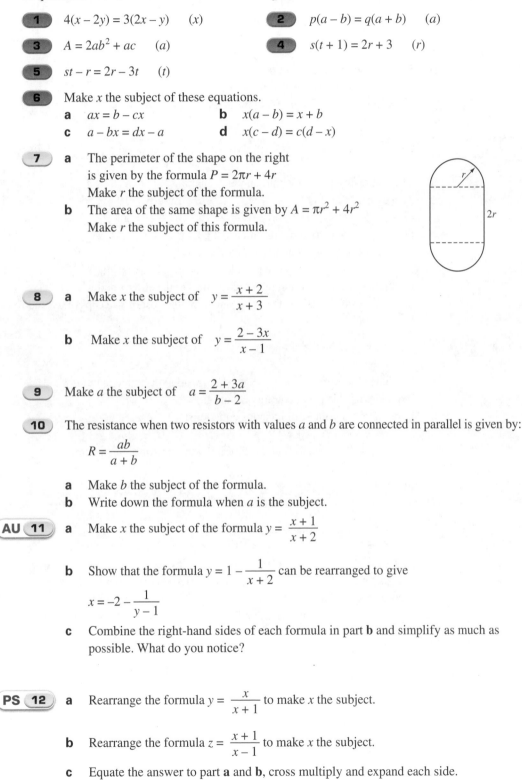

8 **a** Make x the subject of $y = \dfrac{x + 2}{x + 3}$

    b Make x the subject of $y = \dfrac{2 - 3x}{x - 1}$

9 Make a the subject of $a = \dfrac{2 + 3a}{b - 2}$

10 The resistance when two resistors with values a and b are connected in parallel is given by:

$$R = \dfrac{ab}{a + b}$$

    a Make b the subject of the formula.
    b Write down the formula when a is the subject.

AU 11 **a** Make x the subject of the formula $y = \dfrac{x + 1}{x + 2}$

    b Show that the formula $y = 1 - \dfrac{1}{x + 2}$ can be rearranged to give

    $x = -2 - \dfrac{1}{y - 1}$

    c Combine the right-hand sides of each formula in part **b** and simplify as much as possible. What do you notice?

PS 12 **a** Rearrange the formula $y = \dfrac{x}{x + 1}$ to make x the subject.

    b Rearrange the formula $z = \dfrac{x + 1}{x - 1}$ to make x the subject.

    c Equate the answer to part **a** and **b**, cross multiply and expand each side.
    Make y the subject of the resultant formula.

12.3 Simultaneous equations

HOMEWORK 12C

PS 1 In this sequence, the next term is found by multiplying the previous term by a and then subtracting b. a and b are positive whole numbers.

  5   23   113   ....

  a Explain why $5a - b = 23$.

  b Set up another equation in a and b.

  c Solve the equations to solve for a and b.

  d Work out the next two terms in the sequence.

HOMEWORK 12D

1 Solve the following pairs of simultaneous equations.

  a $3x + 2y = 12$ **b** $4x + 3y = 37$ **c** $2x + 3y = 19$ **d** $5x - 2y = 14$
  $4x - y = 5$ $2x + y = 17$ $6x + 2y = 22$ $3x - y = 9$

  e Four cups of tea and three biscuits cost £3.35.

  Three cups of tea and one biscuit cost £2.20

  Let x be the cost of a cup of tea and y be the cost of a biscuit.

   i Set up a pair of simultaneous equations connecting x and y.

   ii Solve your equations for x and y and find the cost of five cups of tea and four biscuits.

AU 2 **a** Tommy is solving the simultaneous equations:

  $5x - y = 9$ and $15x - 3y = 27$.

  He finds a solution of $x = 2$, $y = 1$, which works for both equations.

  Explain why this is not a unique solution.

  b Laura is solving the simultaneous equations:

  $5x - y = 9$ and $15x - 3y = 18$.

  Why is it impossible to find a solution that works for both equations?

HOMEWORK 12E

1 Solve the following simultaneous equations.

  a $6x + 5y = 23$ **b** $3x - 4y = 13$ **c** $8x - 2y = 14$ **d** $5x + 2y = 33$
  $5x + 3y = 18$ $2x + 3y = 20$ $6x + 4y = 27$ $4x + 5y = 23$

  e It costs two adults and three children £28.50 to go to the cinema.

  It costs three adults and two children £31.50 to go to the cinema.

  Let the price of an adult ticket be £x and the price of a child's ticket be £y.

   i Set up a pair of simultaneous equations connecting x and y.

   ii Solve your equations for x and y.

PS 2 Here are four equations.

  A: $2x - y = 8$

  B: $3x + y = 19$

  C: $4x + y = 16$

  D: $3x + 2y = 7$

Here are four sets of (x, y) values:

  $(5.4, 2.8)$ $(4, 0)$ $(-3, 28)$ $(5, -4)$

Match each pair of (x, y) values to a **pair** of equations.

12.4 Solving problems using simultaneous equations

Read each situation carefully, then make a pair of simultaneous equations in order to solve the problem.

 1 A book and a CD cost £14.00 together. The CD costs £7 more than the book. How much does each cost?

PS 2 Ten second-class and six first-class stamps cost £4.96.
Eight second-class and 10 first-class stamps cost £5.84.
How much do I pay for three second-class and four first-class stamps?

3 At the shop, Henri pays £4.37 for six cans of cola and five chocolate bars. On his next visit to the shop he pays £2 for three cans of cola and two chocolate bars. A few days later, he wants to buy two cans of cola and a chocolate bar. How much will they cost him?

FM 4 In her storeroom, Chef Mischa has bags of sugar and rice. The bags are not individually marked, but three bags of sugar and four bags of rice weigh 12 kg. Five bags of sugar and two bags of rice weigh 13 kg.
Help Chef Mischa to work out the weight of two bags of sugar and five bags of rice.

FM 5 Ina wants to buy some snacks for her friends. She works out from the labelling that two cakes and three bags of peanuts contain 63 g of fat; one cake and four bags of peanuts contain 64 g of fat. Help her to work out how many grams of fat there are in each item.

PS 6 The difference between my son's age and my age is 28 years.
Five years ago my age was double that of my son.
Let my age now be x and my son's age now be y.
a Explain why $x - 5 = 2(y - 5)$.
b Find the values of x and y.

FM 7 In a record shop, three CDs and five DVDs cost £77.50.
In the same shop, three CDs and three DVDs cost £55.50.
a Using c to represent the cost of a CD and d to represent the cost of a DVD set up the above information as a pair of simultaneous equations.
b Solve the equations.
c Work out the cost of four CDs and six DVDs.

FM 8 Four apples and two oranges cost £2.04.
Five apples and one orange costs £1.71.
Baz buys four apples and eight oranges.
How much change will he get from a £10 note?

FM 9 Wath School buys basic scientific calculators and graphical calculators to sell to students.
An order for 30 basic scientific calculators and 25 graphical calculators came to a total of £1240. Another order for 25 basic scientific calculators and 10 graphical calculators came to a total of £551.25.
Using £x to represent the cost of basic scientific calculators and £y to represent the cost of graphical calculators, set up and solve a pair of simultaneous equations to find the cost of the next order, for 35 basic scientific calculators and 15 graphical calculators.

FM 10 Five bags of compost and four bags of pebbles weigh 340 kg.

Three bags of compost and five bags of pebbles weigh 321 kg.

Carol wants six bags of compost and eight bags of pebbles.

Her trailer has a safe working load of 500 kg.

Can Carol carry all the bags safely on her trailer?

12.5 Linear and non-linear simultaneous equations

HOMEWORK 12G

Solve these pairs of simultaneous equations.

1 $xy = 3$
$y = x - 2$

2 $xy = 2$
$2y - x = 3$

3 $x^2 + y^2 = 29$
$x - y = 7$

4 $y = x^2 + 3x + 4$
$y = 1 - x$

5 $y = x^2 - 3x + 5$
$y = 2x - 1$

6 $y = 4x^2 + 2x + 1$
$y = 3x^2 + 2x + 2$

PS 7 **a** Solve the simultaneous equations $y = x^2 + 2x - 5$ and $y = 6x - 9$

b Which of the sketches below represents the graph of the equations in **a**?
Explain your choice.

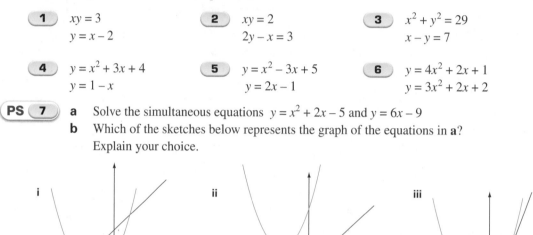

i   ii   iii

AU 8 The solutions of the simultaneous equations $y = x^2 - 3x - 3$ and $y = ax + b$ are (6, 15) and (−1, 1).

Find the values of a and b.

12.6 Algebraic fractions

HOMEWORK 12H

1 Simplify each of these.

a $\dfrac{2x}{3} + \dfrac{4x}{5}$

b $\dfrac{x+1}{3} + \dfrac{x+3}{2}$

c $\dfrac{2x-3}{2} + \dfrac{5x-1}{3}$

2 Simplify each of these.

a $\dfrac{3x}{4} - \dfrac{2x}{5}$

b $\dfrac{x+2}{2} - \dfrac{x+1}{5}$

c $\dfrac{4x-1}{2} - \dfrac{2x-4}{3}$

3 Solve the following equations.

a $\dfrac{2x}{3} + \dfrac{4x}{5} = 11$

b $\dfrac{x+1}{3} + \dfrac{x+3}{2} = 10$

c $\dfrac{2x-5}{2} - \dfrac{x-1}{3} = 1$

4 Simplify each of these.

a $\dfrac{3x}{2} \times \dfrac{4x}{5}$ b $\dfrac{x+1}{4} \times \dfrac{3}{2x+2}$ c $\dfrac{2x-1}{2} \times \dfrac{4}{3x-1}$

5 Simplify each of these.

a $\dfrac{x}{4} \div \dfrac{2x}{5}$ b $\dfrac{x+3}{2} \div \dfrac{2x+6}{5}$ c $\dfrac{4x-2}{3} \div \dfrac{2x-1}{4}$

6 Show that $\dfrac{3}{x+2} + \dfrac{5}{2x-1} = 2$ simplifies to $4x^2 - 5x - 11 = 0$

7 Solve the following equations.

a $\dfrac{3}{x-1} + \dfrac{2}{2x+3} = 5$ b $\dfrac{5}{3x+2} - \dfrac{3}{2x-3} = 4$

8 Simplify the expression $\dfrac{x^2 - 2x - 3}{2x^2 - 10x + 12}$

 9 For homework a teacher asks her class to simplify the expression $\dfrac{x^2 - 9}{x^2 + 2x - 3}$

This is Tim's answer: $\dfrac{x^2 - 9}{x^2 + 2x - 3}$

$$\dfrac{x^2 - \cancel{9}^{\,3}}{\cancel{x}(x+2) - \cancel{3}_1}$$

$$= \dfrac{x-3}{x+2-1} = \dfrac{x-3}{x-1}$$

When she marked the homework, the teacher was in a hurry and only checked the answer, which was correct.

However, Tim made several mistakes.

What are they?

AU 10 An expression of the form $\dfrac{ax^2 - b}{cx^2 + dx^2 - 1}$ simplifies to $\dfrac{3x+4}{x+2}$

What was the original expression?

12.7 Algebraic proof

HOMEWORK 12I

1 If m and n are integers then $(m^2 - n^2)$, $2mn$ and $(m^2 + n^2)$ will form three sides of a right-angled triangle. e.g. Let $m = 5$ and $n = 3$, $m^2 - n^2 = 16$, $2mn = 30$, $m^2 + n^2 = 34$ and $34^2 = 1156$, $16^2 + 30^2 = 256 + 900 = 1156$

Prove this result.

2 Explain why the triangular number sequence 1, 3, 6, 10, 15, 21, 28, ... follows the pattern of two odd numbers followed by two even numbers.

3 $10p + q$ is a multiple of 7. Prove that $3p + q$ is also a multiple of 7.

4   **a**   Show that $2(5(x-2)+y) = 10(x-1)+2y-10$.

       b   Prove that this trick works:

           Think of two numbers less than 10.

           Subtract 2 from the larger number and then multiply by 5.

           Add the smaller number and multiply by 2.

           Add 9 and subtract the smaller number.

           Add 1 to both the tens digits and the units digits to obtain the numbers first thought of.

       c   Prove why the following trick works.

           Choose two numbers. One with one digit, the other with two digits.

           Subtract 9 times the first number from 10 times the second number.

           The units digit of the answer is the single digit number chosen and the sum of the other digits plus the units digit is the other number chosen.

           e.g. Choose 7 and 23. $(23 \times 10) - (7 \times 9) = 167$

           The single digit number chosen is 7, the two-digit number chosen is $16 + 7 = 23$

5   Prove that: $(3n-1)^2 + (3n)^2 + (3n+1)^2 = (5n)^2 + (n-1)^2 + (n+1)^2$

6   Prove that: $(n-1)^2 + n^2 + (n+1)^2 = 3n^2 + 2$

7   **a**   What is the mth term of the sequence 4, 9, 14, 19, 24, … ?

       b   What is the mth term of the sequence 5, 10, 15, 20, 25, … ?

       c   If T_n represent the nth triangular number, prove that $T_n = \frac{1}{2}n(n+1)$

       d   Prove that T_n is a multiple of 5, when n is a member of the series

           4, 5, 14, 15, 19, 20, 24, 25, …

HOMEWORK 12J

1   If T_n is any triangular number, prove that:

       $3T_n = T_{2n+1} - T_{n+1}$ e.g. $T_4 = 10$, $T_9 = 45$, $T_5 = 15$; $3 \times 10 = 45 - 15$

2   If T_n is any triangular number, prove that $\dfrac{T_n - 1}{(T_n)} = \dfrac{(n-1)(n+2)}{n(n+1)}$

3   **a**   What is the nth term of the sequence 1, 4, 7, 10, 13, 16, … .

       b   Explain why there is no multiple of 3 in the sequence.

       c   Prove that the sum of any 5 consecutive numbers in the sequence is a multiple of 5.

4   This question was first set in an examination in 1929.

       $10^x = \dfrac{a}{b}$, $10^y = \dfrac{b}{a}$, Prove that $x + y = 0$

5 m and n are integers. Explain why:

       a   the product $n(n+1)$ must be even   **b** $2m + 1$ is always an odd number

       c   Look at the following numbers pattern.

           $1^2 - 1 = \mathbf{0}$

           $2^2 - 1 = 3$

           $3^2 - 1 = \mathbf{8}$

           $4^2 - 1 = 15$

           $5^2 - 1 = \mathbf{24}$

           $6^2 - 1 = 35$

         i   Extend the pattern for five more lines to show that alternate values are multiples of 8.

        ii   Prove that this is true.

6 a, b, c and d are four consecutive integers. Prove that:

      a $bc - ad = 2$ **b** $ab + bc + cd + da + 1$ is a square number.

7   Prove that the square of the sum of two consecutive integers minus the sum of the squares of the two integers is four times a triangular number.

      e.g. Let the two integers be 6 and 7. $(6 + 7)^2 - (6^2 + 7^2) = 169 - 85 = 84 = 4 \times 21$

8 a, b, c, d are consecutive integers. Prove that $bd - ac$ is always odd.

9 p, q and r are three consecutive numbers. Prove that $pr = q^2 - 1$.

Problem-solving Activity

Algebra and quadrilaterals

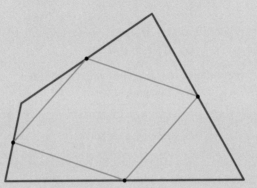

Draw a quadrilateral.

Find the midpoint of each side.

Join the midpoints in order to make a new quadrilateral.

1   Does your new quadrilateral look like a parallelogram? Make appropriate measurements to check whether it could be.

2   Draw some other quadrilaterals and repeat this procedure. Does it look as if you will always get a parallelogram?

3   Can you prove that you always get a parallelogram when you join the midpoints of the sides of a quadrilateral?

13 Number: Variation

13.1 Direct variation

HOMEWORK 13A

For Questions **1** to **5**, first find k, the constant of proportionality, and then the formula connecting the variables.

1 T is directly proportional to M. If $T = 30$ when $M = 5$, find:
   a T when $M = 4$ **b** M when $T = 75$

2 W is directly proportional to F. If $W = 54$ when $F = 3$, find:
   a W when $F = 4$ **b** F when $W = 90$

3 P is directly proportional to A. If $P = 50$ when $A = 2$, find:
   a P when $A = 5$ **b** A when $P = 150$

4 A is directly proportional to t. If $A = 45$ when $t = 5$, find:
   a A when $t = 8$ **b** t when $A = 18$

5 Q varies directly with P. If $Q = 200$ when $P = 5$, find:
   a Q when $P = 3$ **b** P when $Q = 300$

FM 6 The distance covered by a train is directly proportional to the time taken.
The train travels 135 miles in 3 hours.
   a   What distance will the train cover in 4 hours?
   b   What time will it take for the train to cover 315 miles?

FM 7 The cost of petrol is directly proportional to the amount put in the tank.
When 40 litres is used, it costs £32.00. How much:
   a   will it cost for 30 litres?
   b   petrol would there be if the cost were £38.40?
   c   A tank holds 60 litres when full. Petrol is put into the tank until it is full.
       The petrol costs £25.
       How much petrol was in the tank before it was filled up?

FM 8 The number of people who can meet safely in a room is directly proportional to the area of the room. A room with an area of 200 m^2 is safe for 50 people.
   a   How many people can safely meet in a room of area 152 m^2?
   b   A committee has 24 members. What is the smallest room area in which they could safely meet?
   c   An extension is to be built on to a room which is safe for 50 people so that it can accommodate another 20 people.
       The cost of extending is estimated at £160 per square metre.
       How much is the estimate for the extension?

AU 9 A man lays 36 paving stones in 3 hours.
   a   Working at the same rate, how long would he take to lay 45 paving stones?
   b   He works for 7 hours each day. He has 320 stones to lay.
       He employs another worker who can lay 10 stones each hour.
       Will they be able to complete the work in 2 days?

FM Functional Maths   **AU** (AO2) Assessing Understanding   **PS** (AO3) Problem Solving

HOMEWORK 13B

For Questions **1** to **5**, first find k, the constant of proportionality, and then the formula connecting the variables.

1 T is directly proportional to x^2. If $T = 40$ when $x = 2$, find:
   a T when $x = 5$ **b** x when $T = 400$

2 W is directly proportional to M^2. If $W = 10$ when $M = 5$, find:
   a W when $M = 4$ **b** M when $W = 64$

3 A is directly proportional to r^2. If $A = 96$ when $r = 4$, find:
   a A when $r = 5$ **b** r when $A = 12$

4 E varies directly with $\sqrt{C}$. If $E = 60$ when $C = 36$, find:
   a E when $C = 49$ **b** C when $E = 160$

5 X is directly proportional to $\sqrt{Y}$. If $X = 80$ when $Y = 16$, find:
   a X when $Y = 100$ **b** Y when $X = 48$

FM 6 The distance covered by a train is directly proportional to the time taken.
The train travels 144 miles in 4 hours.
   a What distance will the train cover in 2 hours 30 minutes?
   b What time will it take for the train to cover 54 miles?

FM 7 The cost of petrol is directly proportional to the amount put in the tank.
When 40 litres is used, it costs £44.00. How much:
   a will it cost for 18 litres   **b** petrol would there be if the cost were £27.50?

8 y is directly proportional to $\sqrt[3]{x}$.
If $y = 4$ when $x = 8$, find the following:
   a y when $x = 1$
   b x when $y = 250$

FM 9 The number of people who can meet safely in a room is directly proportional to the area of the room. A room with an area of 200 m^2 is safe for 50 people.
   a How many people can safely meet in a room of area 168 m^2?
   b A committee has 36 members. What is the smallest room area in which they could safely meet?

10 A man lays 36 paving stones in 3 hours.
Working at the same rate how long would he take to lay 54 paving stones?

FM 11 An artist is painting pictures.
The amount of time taken to complete a picture is directly proportional to the square of the width of the picture.
   A picture is 30 cm wide and takes 20 hours to complete.
   A buyer wants a picture that is 50 cm wide within 10 days.
   If the artist paints for 6 hours each day, can he complete the picture on time?

A

PS AU **12** Here are three proportion statements and two tables.

a $y \propto x^2$ b $y \propto x$ c $y \propto \sqrt{x}$

A

x	1	4	9
y	2	4	6

B

x	1	2	3
y	4	8	12

Match each table to the correct proportion statement.

13.2 Inverse variation

HOMEWORK 13C

For Questions **1** to **7**, first find the formula connecting the variables.

1 T is inversely proportional to m. If $T = 7$ when $m = 4$, find:

    a T when $m = 5$ **b** m when $T = 56$

2 W is inversely proportional to x. If $W = 6$ when $x = 15$, find:

    a W when $x = 3$ **b** x when $W = 10$

3 M varies inversely with t^2. If $M = 10$ when $t = 2$, find:

    a M when $t = 4$ **b** t when $M = 160$

4 C is inversely proportional to f^2. If $C = 20$ when $f = 3$, find:

    a C when $f = 5$ **b** f when $C = 720$

5 W is inversely proportional to $\sqrt{T}$. If $W = 8$ when $T = 36$, find:

    a W when $T = 25$ **b** T when $W = 0.75$

6 H varies inversely with $\sqrt{g}$. If $H = 20$ when $g = 16$, find:

    a H when $g = 1.25$ **b** g when $H = 40$

7 y is inversely proportional to the cube of x. If $y = 10$ when $x = 1$, find the following:

    a y when $x = 2$

    b x when $y = 270$

FM **8** The brightness of a bulb decreases inversely with the square of the distance away from the bulb. The brightness is 5 candle power at a distance of 10 m.

What is the brightness at a distance of 5 m?

AU **9** In the table, y is inversely proportional to x.

x	2	4	16
y	8		

Copy and complete the table.

10 The density of a series of spheres with the same weight is inversely proportional to the cube of the radius. A sphere with a density of 10 g/cm³ has a radius of 5 cm.

    a What would be the density of a sphere with a radius of 10 cm?

    b If the density was 80 g/cm³ what would the radius of the sphere be?

11 Given that y is inversely proportional to the square of x, and that y is 12 when $x = 4$:

    a find an expression for y in terms of x.

    b Calculate: **i** y when $x = 6$ **ii** x when $y = 36$

FM 12 The time taken to build an extension is inversely proportional to the number of workers. It takes 2 workers 7 days to complete an extension.

a   Three workers start an extension on Monday morning.
    Will they complete it by Friday?
    Show your working.

b   Give a reason why the time taken might not be inversely proportional to the number of workers when the number of workers is very large.

Functional Maths Activity

Taxi fares

Imagine you run a taxi company.

A rival taxi company has a fares structure based on three things:

- a minimum fare of £2.50
- a charge for the time taken
- a charge for the distance travelled.

The charge for the time taken is directly proportional to the time taken.

The charge for the distance travelled is directly proportional to the distance travelled.

You have collected this information about the rival company's fares.

Use this information to work out the charge per mile and the charge per minute, then suggest a competitive pricing structure for your own taxi firm.

Time taken	2 minutes	5 minutes	10 minutes	12 minutes	15 minutes
Distance	1 mile	2 miles	3 miles	5 miles	6 miles
Total charge	£2.50 (minimum fare)	£4.00	£6.50	£9.90	£12.00

14 Number: Number and limits of accuracy

14.1 Limits of accuracy

1 Write down the limits of accuracy of the following.
- **a** 5 cm to the nearest centimetre
- **b** 40 mph to the nearest 10 mph
- **c** 15.2 kg to the nearest tenth of a kilogram
- **d** 75 km to the nearest 5 km

2 Write down the limits of accuracy for each of the following values which are to the given degree of accuracy.

a	7 cm (1 sf)	**b**	18 kg (2 sf)	**c**	30 min (2 sf)
d	747 km (3 sf)	**e**	9.8 m (1 dp)	**f**	32.1 kg (1 dp)
g	3.0 h (1 dp)	**h**	90 g (2 sf)	**i**	4.20 mm (2 dp)
j	2.00 kg (2 dp)	**k**	34.57 min (2 dp)	**l**	100 m (2 sf)

3 Round off these numbers to the degree of accuracy given.
- **a** 45.678 to 3 sf   **b** 19.96 to 2 sf   **c** 0.3213 to 2 dp

4 Write down the upper and lower bounds of each of these values given to the accuracy stated.

a	6 m (1 sf)	**b**	34 kg (2 sf)	**c**	56 min (2 sf)	**d**	80 g (2 sf)
e	3.70 m (2 dp)	**f**	0.9 kg (1 dp)	**g**	0.08 s (2 dp)	**h**	900 g (2 sf)
i	0.70 m (2 dp)	**j**	360 d (3 sf)	**k**	17 weeks (2 sf)	**l**	200 g (2 sf)

 5 A theatre has 365 seats.

For a show, 280 tickets are sold in advance.

The theatre's manager estimates that another 100 people to the nearest 10 will turn up without tickets.

Is it possible they will all get a seat, assuming that 5% of those with tickets do not turn up?

Show clearly how you decide.

AU 6 A parking space is 4.8 metres long to the nearest tenth of a metre.

A car is 4.5 metres long to the nearest half a metre.

Which of the following statements is definitely true?

A: The space is big enough

B: The space is not big enough

C: It is impossible to tell whether or not the space is big enough

Explain how you decide.

AU 7 1 litre = 100 cl

A carton contains 1 litre of milk to the nearest 10 cl.

What is the least amount of milk likely to be in the carton?

Give your answer in centilitres.

8 Billy has 20 identical bricks. Each brick is 15 cm long measured to the nearest centimetre.

   a   What is the greatest length of one brick?

   b   What is the smallest length of one brick?

   c   If the bricks are put end to end, what is the greatest possible length of all the bricks?

   d   If the bricks are put end to end, what is the least possible length of all the bricks?

14.2 Problems involving limits of accuracy

HOMEWORK 14B

1 Cans have a mass of 250 grams to the nearest 10 grams.
What is the minimum and maximum mass of 10 of these cans?

FM 2 The cans in Question **1** are stacked on a shelf.
It can safely hold 15 kg of cans.
What is the maximum number of cans that can safely be put on the shelf?

PS 3 A crate of the cans in Question **1** has a mass of 24 kg.
How many cans could be in the crate?

4 For each of these rectangles, find the limits of accuracy of the area. The accuracy of each measurement is given.

   a   3 cm × 8 cm   (nearest cm)   **b**   3.2 cm × 6.4 cm   (1 dp)

   c   7.86 cm × 18.78 cm   (2 dp)

5 A rectangular garden has sides of 8 m and 5 m, measured to the nearest metre.

   a   Write down the limits of accuracy for each length.

   b   What is the maximum area of the garden?

   c   What is the minimum perimeter of the garden?

6 A playground is measured as 32 m by 45 m, to the nearest metre. Calculate the limits of accuracy for the area of the playground.

7 The measurements of a box are given as 12 cm × 8 cm × 5 cm to the nearest centimetre.
Calculate the limits of accuracy for the volume of the box.

AU 8 Mr Leake is a plumber.
He has a 10-metre piece of pipe correct to the nearest metre.
He uses 2 metres of pipe on his first job, 3 metres on his second job and 4 metres on his third job. Each measurement is to the nearest half a metre.
What is the longest length of pipe he could have left?

AU 9 Belinda is doing a five-mile walk.
She is walking at an average speed of 3 mph to the nearest whole number of miles per hour.
She sets off at 2 pm.
What is the latest time that she will complete the walk?

10 The area of a field is given as 400 m^2, to the nearest 10 m^2. One length is given as 24 m, to the nearest metre. Find the limits of accuracy for the other length of the field.

11 In triangle ABC, AB = 8 cm, BC = 6 cm, and ABC = 42°. All the measurements are given to the nearest unit.
Calculate the limits of accuracy for the area of the triangle.

12 A stopwatch records the time for the winner of a 100-metre race as 12.3 seconds, measured to the nearest one-tenth of a second.

   a What are the greatest and least possible times for the winner?

   b The length of the 100-metre track is correct to the nearest 1 centimetre. What are the greatest and least possible lengths of the track?

   c What is the fastest possible average speed of the winner?

13 A cube has a volume of 27 cm^3, to the nearest cm^3. Find the range of possible values of the side length of the cube.

14 A cube has a volume of 125 cm^3, to the nearest 1 cm^3. Find the limits of accuracy of the area of one side of the square base.

PS 15 In the triangle ABC, the length of side AB is 42 cm to the nearest centimetre. The length of side AC is 35 cm to the nearest centimetre. The angle C is 61° to the nearest degree. What is the largest possible size that angle B could be?

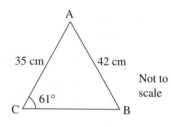

Not to scale

HINTS AND TIPS

You will need to use the Sine Rule.

Functional Maths Activity

Lorries and loads

You are the manager of a haulage company. You own a lorry with 6 axles, with a maximum axle weight limit of 10.5 tonnes. You also own a lorry with 5 axles, with a maximum axle weight limit of 11.5 tonnes. This lorry is 10% cheaper to run per trip than the 6-axle lorry.

These are the load restrictions for heavy goods vehicles.

Type of lorry	Load limit
6 axles, max. axle weight limit 10.5 tonnes	44 tonnes
5–6 axles, max. axle weight limit 11.5 tonnes	40 tonnes

You want to deliver pallets of goods.
Each pallet weighs 500 kg to the nearest 50 kg.

Task 1
A company orders 80 pallets.
Which lorry would you use for this job?
Show clearly any decisions you make and explain how you decide.

Task 2
A company orders 150 pallets.
Which lorry would you use for this job?
Show clearly any decisions you make and explain how you decide.

Task 3
A company orders 159 pallets.
Which lorry would you use for this job?
Show clearly any decisions you make and explain how you decide.

Task 4
Devise a planning chart for orders up to 250 pallets to help you decide which lorry to use.

15.1 Properties of vectors

HOMEWORK 15A

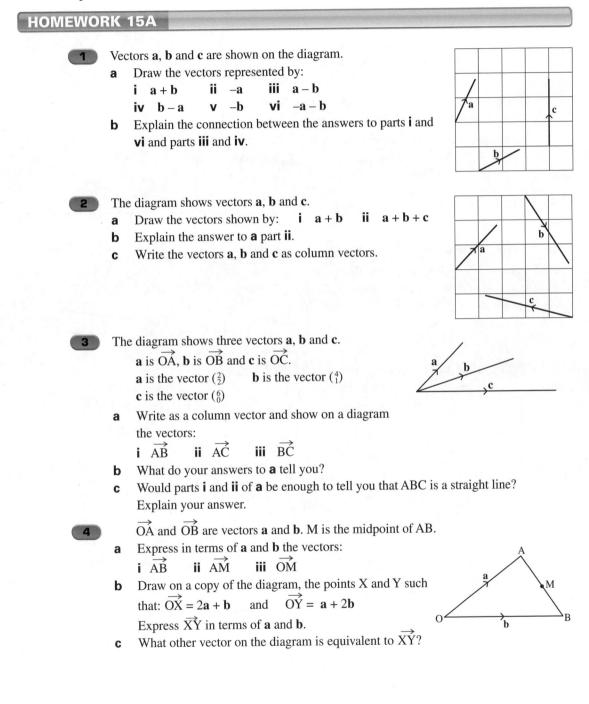

1 Vectors **a**, **b** and **c** are shown on the diagram.
  a Draw the vectors represented by:
   i a + b   **ii** –a   **iii** a – b
   iv b – a   **v** –b   **vi** –a – b
  b Explain the connection between the answers to parts **i** and
   vi and parts **iii** and **iv**.

2 The diagram shows vectors **a**, **b** and **c**.
  a Draw the vectors shown by: **i** a + b   **ii** a + b + c
  b Explain the answer to **a** part **ii**.
  c Write the vectors **a**, **b** and **c** as column vectors.

3 The diagram shows three vectors **a**, **b** and **c**.
   a is $\overrightarrow{OA}$, **b** is $\overrightarrow{OB}$ and **c** is $\overrightarrow{OC}$.
   a is the vector $\binom{2}{2}$ **b** is the vector $\binom{4}{1}$
   c is the vector $\binom{6}{0}$
  a Write as a column vector and show on a diagram
   the vectors:
   i $\overrightarrow{AB}$ **ii** $\overrightarrow{AC}$ **iii** $\overrightarrow{BC}$
  b What do your answers to **a** tell you?
  c Would parts **i** and **ii** of **a** be enough to tell you that ABC is a straight line?
   Explain your answer.

4 $\overrightarrow{OA}$ and $\overrightarrow{OB}$ are vectors **a** and **b**. M is the midpoint of AB.
  a Express in terms of **a** and **b** the vectors:
   i $\overrightarrow{AB}$ **ii** $\overrightarrow{AM}$ **iii** $\overrightarrow{OM}$
  b Draw on a copy of the diagram, the points X and Y such
   that: $\overrightarrow{OX} = 2\mathbf{a} + \mathbf{b}$ and $\overrightarrow{OY} = \mathbf{a} + 2\mathbf{b}$
   Express $\overrightarrow{XY}$ in terms of **a** and **b**.
  c What other vector on the diagram is equivalent to $\overrightarrow{XY}$?

5 OACB is a trapezium, where $\overrightarrow{OA}$ = **a**, $\overrightarrow{OB}$ = **b** and $\overrightarrow{BC}$ = 2**a**. P and Q are the midpoints of $\overrightarrow{OB}$ and $\overrightarrow{AC}$. Express in terms of **a** and **b**.

a $\overrightarrow{OP}$ **b** $\overrightarrow{AQ}$ **c** $\overrightarrow{PQ}$ **d** How can you tell that $\overrightarrow{PQ}$ is parallel to $\overrightarrow{OA}$?

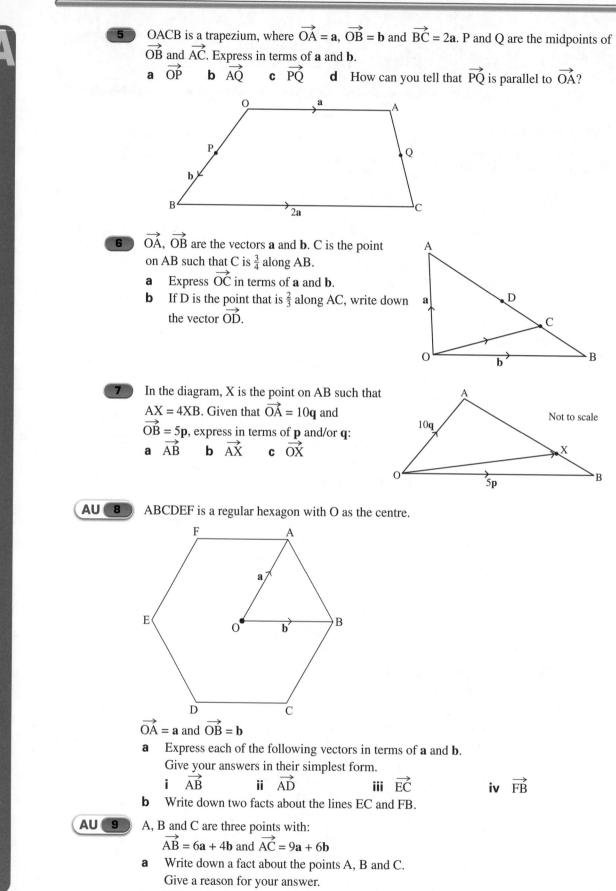

6 $\overrightarrow{OA}$, $\overrightarrow{OB}$ are the vectors **a** and **b**. C is the point on AB such that C is $\frac{3}{4}$ along AB.

a Express $\overrightarrow{OC}$ in terms of **a** and **b**.

b If D is the point that is $\frac{2}{3}$ along AC, write down the vector $\overrightarrow{OD}$.

7 In the diagram, X is the point on AB such that AX = 4XB. Given that $\overrightarrow{OA}$ = 10**q** and $\overrightarrow{OB}$ = 5**p**, express in terms of **p** and/or **q**:

a $\overrightarrow{AB}$ **b** $\overrightarrow{AX}$ **c** $\overrightarrow{OX}$

Not to scale

AU 8 ABCDEF is a regular hexagon with O as the centre.

$\overrightarrow{OA}$ = **a** and $\overrightarrow{OB}$ = **b**

a Express each of the following vectors in terms of **a** and **b**.
Give your answers in their simplest form.

i $\overrightarrow{AB}$ **ii** $\overrightarrow{AD}$ **iii** $\overrightarrow{EC}$ **iv** $\overrightarrow{FB}$

b Write down two facts about the lines EC and FB.

AU 9 A, B and C are three points with:

$\overrightarrow{AB}$ = 6**a** + 4**b** and $\overrightarrow{AC}$ = 9**a** + 6**b**

a Write down a fact about the points A, B and C.
Give a reason for your answer.

b Write down the ratio of the lengths AB : BC in its simplest form.

15.2 Vectors in geometry

HOMEWORK 15B

1 OACB is a rectangle. $\overrightarrow{OA} = \mathbf{a}$ and $\overrightarrow{OB} = \mathbf{b}$.
Q is the midpoint of BC and P divides BA in the ratio 1 : 2. Find the vectors:

a $\overrightarrow{BP}$ b $\overrightarrow{OP}$ c $\overrightarrow{OQ}$

d Explain the relationship between O, P and Q.

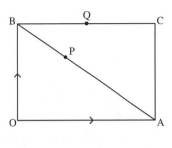

2 $\overrightarrow{OA} = \mathbf{a}$. $\overrightarrow{OB} = \mathbf{b}$. P is the point that divides OB in the ratio 1 : 2. Q is the point that divides OA in the ratio 2 : 1.

a Express in terms of **a** and **b**:   **i** $\overrightarrow{AP}$ **ii** $\overrightarrow{BQ}$

b Explain why OR can be written as $\mathbf{a} + n\overrightarrow{AP}$.

c Explain why OR can be written as $\mathbf{b} + m\overrightarrow{BQ}$.

d Show that the expressions in parts **b** and **c** are equal when $n = \frac{3}{7}$ and $m = \frac{6}{7}$.

e Hence find the vector $\overrightarrow{OR}$ in terms of **a** and **b**.

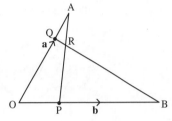

3 OAB is a triangle. $\overrightarrow{OA} = \mathbf{a}$. $\overrightarrow{OB} = \mathbf{b}$. R is the midpoint of AB. Q is the midpoint of OA.
Find in terms of **a** and **b** the vectors:

a $\overrightarrow{OR}$ b $\overrightarrow{QB}$

c G is the point where OR and QB meet.
Explain why $\overrightarrow{OG}$ can be written both as $n\ \overrightarrow{OR}$ and $\frac{1}{2}\mathbf{a} + m(\overrightarrow{QB})$.

d You are given that $m + n = 1$.
Find values of m and n that satisfy the equations in **c**.

e Hence express $\overrightarrow{OG}$ in terms of **a** and **b**.

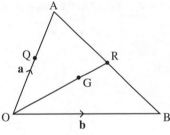

4 OAB is a triangle. P, Q and R are the midpoints of OB, OA and AB. $\overrightarrow{OR} = \mathbf{a}$ and $\overrightarrow{OP} = \mathbf{b}$.

a Express in terms of **a** and **b** the vectors:

i $\overrightarrow{RP}$ **ii** $\overrightarrow{AB}$

b What can you say about $\overrightarrow{RP}$ and $\overrightarrow{AB}$?

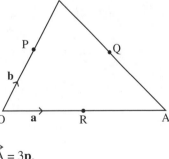

5 OAC is a triangle such that $\overrightarrow{OC} = 12\mathbf{q}$, $OB = 3\mathbf{q}$ and $\overrightarrow{OA} = 3\mathbf{p}$.

a Find in terms of **p** and **q** the vectors:

i $\overrightarrow{AB}$ **ii** $\overrightarrow{AC}$

b Given that $AQ = \frac{1}{3}AC$ express $\overrightarrow{OQ}$ in terms of **p** and **q**.

c Given that $\overrightarrow{OR} = \mathbf{p} + 2\mathbf{q}$ what can you say about the points O, R and Q?

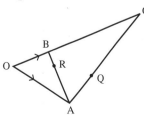

A*

AU 6 OABC is a quadrilateral. $\vec{OA} = \mathbf{a}$, $\vec{OB} = \mathbf{b}$, $\vec{OC} = \mathbf{c}$.
M, N, Q and P are the midpoints of OA, OB, CB and AC, respectively.

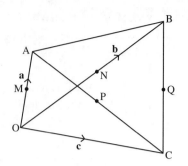

a Find in terms of **a**, **b** and **c** the vectors:

   i $\vec{BC}$ **ii** $\vec{NQ}$ **iii** $\vec{MP}$

b What type of quadrilateral is MNQP?
Explain your answer.

7 OACB and OBRS are parallelograms.
$\vec{OA}$ is **a**, $\vec{OB}$ is **b**, and $\vec{BR}$ is **r**. Find in terms of **a**, **b** and **r** expressions in their simplest forms for:

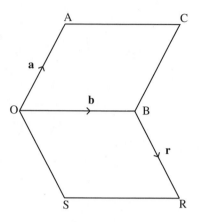

a $\vec{OR}$

b $\vec{SB}$

c $\vec{OX}$, where X is the midpoint of AR,
   i.e. $\frac{1}{2}(\vec{AO} + \vec{OR})$

AU 8 On the diagram, $\vec{OA} = \mathbf{a}$, $\vec{OB} = \mathbf{b}$ and $\vec{OC} = 3\mathbf{b} - 2\mathbf{a}$.

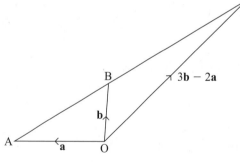

Prove that ABC is a straight line.

AU 9 OACB is a trapezium with $\vec{OA} = 3\mathbf{a}$,
$\vec{OB} = 4\mathbf{b}$ and $\vec{BC} = 6\mathbf{c}$.

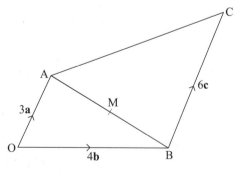

a Find these vectors in terms of **a** and **b**.

   i $\vec{OC}$ **ii** $\vec{AB}$

b Point M lies on AB with
AM : MB = 1 : 2.
Find $\vec{OM}$ in terms of **a** and **b**.
Give your answer in its simplest form.

c Explain why OMC is a straight line.

15.3 Geometric proof

HOMEWORK 15C

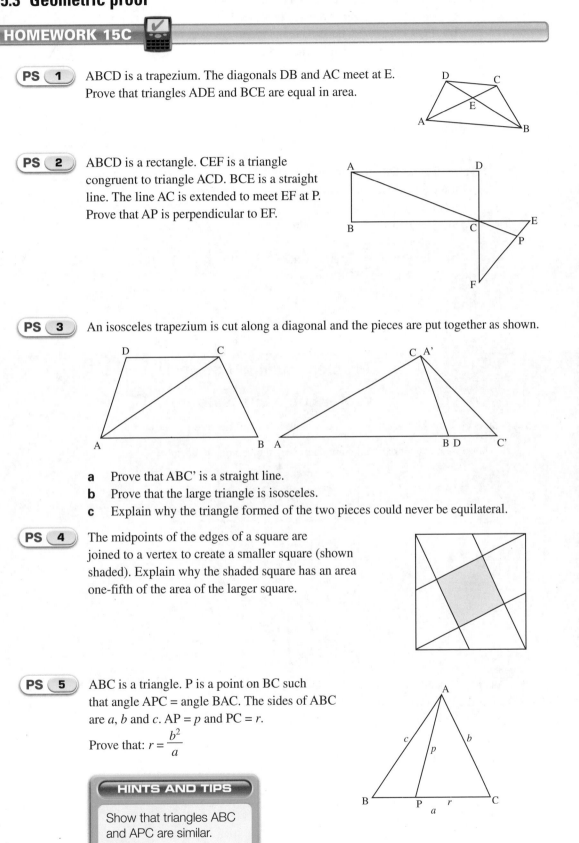

PS 1 ABCD is a trapezium. The diagonals DB and AC meet at E. Prove that triangles ADE and BCE are equal in area.

PS 2 ABCD is a rectangle. CEF is a triangle congruent to triangle ACD. BCE is a straight line. The line AC is extended to meet EF at P. Prove that AP is perpendicular to EF.

PS 3 An isosceles trapezium is cut along a diagonal and the pieces are put together as shown.

a Prove that ABC' is a straight line.
b Prove that the large triangle is isosceles.
c Explain why the triangle formed of the two pieces could never be equilateral.

PS 4 The midpoints of the edges of a square are joined to a vertex to create a smaller square (shown shaded). Explain why the shaded square has an area one-fifth of the area of the larger square.

PS 5 ABC is a triangle. P is a point on BC such that angle APC = angle BAC. The sides of ABC are a, b and c. AP = p and PC = r.

Prove that: $r = \dfrac{b^2}{a}$

HINTS AND TIPS

Show that triangles ABC and APC are similar.

A*

PS 6

a Prove that the angles subtended by a chord at the circumference of a circle are equal.

b PQRS is a cyclic quadrilateral. PR and QS meet at T. Angles x, $2x$, $3x$ and $5x$ are marked on the diagram.

   i Find x.

   ii Show that the angles of the quadrilateral and angle STP form a number sequence.

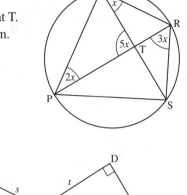

PS 7 ACB and ADB are right-angled triangles. The lengths are as marked.

a Use Pythagoras' theorem to show that $x^2 = r^2 + s^2$.

b Use Pythagoras' theorem on both triangles ACB and ADB to prove that: $xt = sy$

Problem-solving Activity

Vectors on a chess board

A Knight is on the white square near the bottom left-hand corner of a chess board.

The two possible moves it can make are shown by the vectors **a** and **b**.

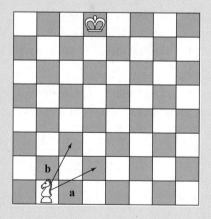

Using only combinations of these two types of move, how many different ways can the Knight reach the King at the top of the chess board?

16.1 Transformations of the graph $y = f(x)$

HOMEWORK 16A

You may use a graphical calculator or a graph drawing program to do this exercise.

1 On the same axes sketch the graphs of:
   a $y = x^2$ **b** $y = 2x^2$ **c** $y = x^2 + 2$ **d** $y = (x + 2)^2$
   e Describe the transformation(s) that take(s) the graph in part **a** to each of the graphs in parts **b** to **d**.

2 On the same axes sketch the graphs of:
   a $y = x^2$ **b** $y = 3x^2 + 2$ **c** $y = x^2 - 3$ **d** $y = \frac{1}{2}x^2 + 1$
   e Describe the transformation(s) that take(s) the graph in part **a** to each of the graphs in parts **b** to **d**.

3 On the same axes sketch the graphs of:
   a $y = x^2$ **b** $y = (x + 4)^2$ **c** $y = -x^2$ **d** $y = 2 - x^2$
   e Describe the transformation(s) that take(s) the graph in part **a** to each of the graphs in parts **b** to **d**.

4 On the same axes sketch the graphs of:
   a $y = \sin x$ **b** $y = 3\sin x$ **c** $y = \sin x + 3$ **d** $y = \sin(x + 30°)$
   e Describe the transformation(s) that take(s) the graph in part **a** to each of the graphs in parts **b** to **d**.

5 On the same axes sketch the graphs of:
   a $y = \sin x$ **b** $y = -\sin x$ **c** $y = \sin\frac{x}{3}$ **d** $y = 3\sin\frac{x}{2}$
   e Describe the transformation(s) that take(s) the graph in part a to each of the graphs in parts **b** to **d**.

6 On the same axes sketch the graphs of:
   a $y = \sin x$ **b** $y = 3\sin x$ **c** $y = \sin(x + 45°)$ **d** $y = 2\sin(x + 90°)$
   e Describe the transformation(s) that take(s) the graph in part **a** to the graphs in parts **b** to **d**.

7 On the same axes sketch the graphs of:
   a $y = \cos x$ **b** $y = -\cos x$ **c** $y = \cos x + 4$ **d** $y = 2\cos x$
   e Describe the transformation(s) that take(s) the graph in part **a** to each of the graphs in parts **b** to **d**.

8 On the same axes sketch the graphs of:
   a $y = \cos x$ **b** $y = 3\cos x$ **c** $y = \cos(x + 60°)$ **d** $y = 2\cos x + 3$
   e Describe the transformation(s) that take(s) the graph in part **a** to each of the graphs in parts **b** to **d**.

9 Explain why the graphs of $y = \cos x$ and $y = \sin(x + 90°)$ are the same.

A*

10 The table shows some values of the function $f(x) = (x - 2)^2 + 4$, where $-3 < x < 4$.

    a Draw the graph of $y = f(x)$.

    b On the same axes, draw the graph of: $y = x^2$.

x	-3	-2	-1	0	1	2	3	4
$f(x)$	29	20	13	8	5	4	5	8

    c Describe how the graph of $y = (x - 2)^2 + 4$ can be obtained from the graph $y = x^2$ by a transformation. State clearly what this transformation is.

PS 11 The graphs below are all transformations of $y = x^2$. Two points through which each graph passes are indicated. Use this information to work out the equation of each graph.

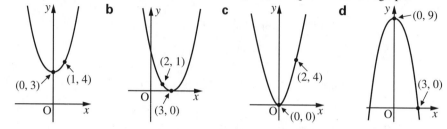

PS 12 The graphs below are all transformations of $y = \sin x$. Two points through which each graph passes are indicated. Use this information to work out the equation of each graph.

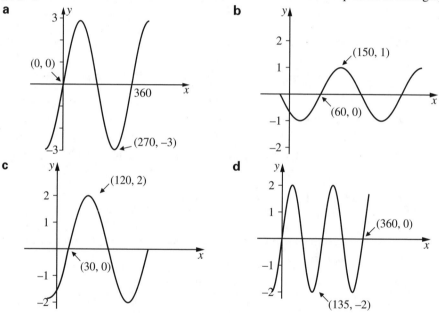

AU 13 Below are the graphs of: $y = -\sin x$ and $y = \cos x$

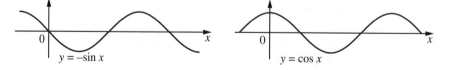

    a Describe a series of transformations that would take the first graph to the second.

    b Which of the following is equivalent to $y = -\sin x$?

       i $y = \sin(x + 180°)$ **ii** $y = \cos(x + 90°)$ **iii** $y = 2\sin\frac{x}{2}$

14 The graph of $y = f(x)$ has been drawn.
Sketch the graphs of:

    a $y = f(x) - 2$ **b** $y = f(x - 2)$

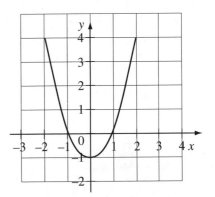

Problem-solving Activity

Transforming graphs

The sketch below shows the graphs of $y = x^2$ and $y = x^2 + 3x$

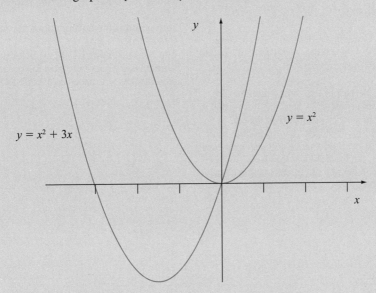

1 **a** Write down the coordinates of the minimum point of the equation:

      $y = x^2 + 3x$

  b Describe how the graph of $y = x^2 + 3x$ can be transformed from the graph of:

      $y = x^2$

2 Now draw graphs of $y = x^2 + 4x$, $y = x^2 - 2x$ and $y = x^2 - x$ on the same axes.

  a Describe anything that you notice about these graphs.

  b Describe how you could obtain the graph of $y = x^2 + ax$ by transforming the graph of $y = x^2$

William Collins' dream of knowledge for all began with the publication of his first book in 1819. A self-educated mill worker, he not only enriched millions of lives, but also founded a flourishing publishing house. Today, staying true to this spirit, Collins books are packed with inspiration, innovation and practical expertise. They place you at the centre of a world of possibility and give you exactly what you need to explore it.

Collins. Freedom to teach.

Published by Collins
An imprint of HarperCollins*Publishers*
77–85 Fulham Palace Road
Hammersmith
London
W6 8JB

> Browse the complete Collins catalogue at
> www.collinseducation.com

© HarperCollins*Publishers* Limited 2010

10 9 8 7 6 5 4

ISBN 978-0-00-734030-9

Brian Speed, Keith Gordon, Keith Evans, Trevor Senior and Chris Pearce assert their moral rights to be identified as the authors of this work

British Library Cataloguing in Publication Data
A Catalogue record for this publication is available from the British Library

Commissioned by Katie Sergeant
Project managed by Patricia Briggs
Edited by Brian Asbury
Answers checked by Steven Matchett and Joan Miller
Cover design by Angela English
Concept design by Nigel Jordan
Illustrations by Wearset Publishing Services
Typesetting by Wearset Publishing Services
Production by Leonie Kellman
Printed and bound by Printing Express, Hong Kong

Important information about the Student Book CD-ROM

The accompanying CD-ROM is for home use only. You cannot copy or save the files to your hard drive and it will work only when placed in the CD-ROM drive.